Marek "Mark" J. Wagner

Magnificent Awareness

The Law of Everything

The Paramount Law of Transformation

Marek "Mark" J. Wagner

The Paramount Law of Transformation

The Law of Everything
Magnificent Awareness

Pierwotne Prawo Transformacji
Uniwersalna Teoria Wszystkiego

Honest Dragon Publishing

All rights reserved.

ISBN: 978-83-937658-3-6

www.zhibit.org/mjw23
www.scribd.com/mjw23
E-mail: q7q7mark@gmail.com

Tekst/Text, Szata Graficzna/Graphic Design/Editing:
Honest Dragon Publishing 2000 / 2014

Images / zdjecia, editing: Honest Dragon Publishing

Please support an Open Letter to the UN and European Union
Proszę o poparcie Listu Otwartego do Narodów Zjednoczonych
Karty Praw Człowieka / Petycji do Unii Europejskiej (Human Rights Act).

http://www.scribd.com/doc/67148949/United-Nations-Human-Rights-Council-European-Human-Rights-Act

All Rights Resreved 2004/2015   Tel. 536 508 394   www.scribd.com/mjw23   Marek „Mark" J. Wagner   www.zhibit.org/mjw23   E-mail: q7q7mark@gmail.com

3

* The Paramount Law of Transformation. Parallel Universes vs Cellular Automation. Universe is within us. How to „Resurrect" the Universe.
* The Paramount Law of Transformation. Projections vs Universal Performance
* The Paramount Law of Transformation. Eloquence & Elegance of Intelligent Design.Fibonacci Sequence vs Primes vs Universal Morphogenesis vs Universal Metabolism. Gravity: Hot Light vs Cold Light.
* The Paramount Law of Transformation. Great Exit Mode vs Procreation of the World.
* The Paramount Law of Transformation vs Biological Blueprint of the Universe. Neutrino. 3I Rule. Particles can learn beyond adaptation … .
* The Paramount Law of Transformation. Biological Blueprint of the Universe. Memory: molecuar-kinetic memory vs interactive memory. Mathematics vs Aesthetic Paradigm. The 8th Cycle of the Creation.
* The Paramount Law of Transformation. Mathematics vs 3D vs Molecules vs Energies vs Reality.
* The Paramount Law of Transformation. Human vs Positive Compounding of Energy vs Negative Compounding Energy.
* The Paramount Law of Transformation. Vortex of Matter vs Resonant Waves vs Initiation Sequence.
* The Paramount Law of Transformation. Biological Blueprint of the Universe. Space Program sinc3 1452. Quantum Data. 3D.Symmetry. Quantum Data: Quantum (the essence):
   Singularity dwell in whole (within) ,yet, in whole dwell in singularity = I am in you, you are in me … .
* The Paramount Law of Transformation. Biological Blueprint of the Universe. Quantum Unity. The Power of Simplicity. Improvisation. 3D Projection.
* The Paramount Law of Transformation. Biological Blueprint of the Universe. (ᾥ) Magnificent Quantum Awareness… .
* The Paramount Law of Transformation. Biological Blueprint of the Universe. Intelligent, transformative function within Intelligent Design is shaping reality. Biological accelerator.
* The Paramount Law of Transformation. Biological Blueprint of the Universe. Universe; individuality vs parallels vs quantified essence. Molecules; formation of physical data.
* The Paramount Law of Transformation. Biological Blueprint of the Universe. Molecules; formation of physical data.
* The Paramount Law of Transformation. Biological Blueprint of the Universe. Universe; Physical data vs Rising of Awareness.
* The Paramount Law of Transformation. Biological Blueprint of the Universe. Space Program sinc3 1452. Rising of Awareness… .
* The Paramount Law of Transformation. Biological Blueprint of the Universe. Space Program sinc3 1452. „Zero" ? Yes … .
* The Paramount Law of Transformation. Biological Blueprint of the Universe. Space Program sinc3 1452. Micro-Cosmos vs Macro-cosmos. Sphere vs Platonic Solids vs Power of Quantified Transformations… .
* The Paramount Law of Transformation. Biological Blueprint of the Universe. Space Program sinc3 1452. Mass. Universal Composition.
* The Paramount Law of Transformation. Biological Blueprint of the Universe. Space Program sinc3 1452. Logic vs Divine Affairs … .
* The Paramount Law of Transformation. Biological Blueprint of the Universe. Space Program since 1452. Science vs Physics vs Love … .
* The Paramount Law of Transformation. Biological Blueprint of the Universe. Space Program sinc3 1452. Energy: High Velocity Plasma … .
* The Paramount Law of Transformation. Biological Blueprint of the Universe. Space Program sinc3 1452. Univ3rse vs Gravity … .
* The Paramount Law of Transformation. Biological Blueprint of the Universe. Space Program sinc3 1452. Univ3rse vs Vortex of Matter … .
* The Paramount Law of Transformation. Biological Blueprint of the Universe. Space Program sinc3 1452. Space Bridges … .
* The Paramount Law of Transformation. Biological Blueprint of the Universe. Space Program sinc3 1452; Simplicity vs Complexity … .
* The Paramount Law of Transformation. Biological Blueprint of the Universe. Space Program sinc3 1452; Red … .
* The Paramount Law of Transformation. Biological Blueprint of the Universe. Space Program sinc3 1452; Flowers vs Electromagnetic Spectrum … .
* The Paramount Law of Transformation. Biological Blueprint of the Universe. Space Program sinc3 1452; Necessity vs Pleasure … .
* The Paramount Law of Transformation. Biological Blueprint of the Universe. Space Program sinc3 1452; Quantified Journey … .
* The Paramount Law of Transformation. Biological Blueprint of the Universe. Space Program sinc3 1452; Engine of Universal Geometry; Shapes … .
* The Paramount Law of Transformation. Biological Blueprint of the Universe. Space Program sinc3 1452; Subatomic Plasma … .
* The Paramount Law of Transformation. B;ological Blueprint of the Universe. Spac3 Program Since 1452. Space Bridges ? Y3s… .Mirroring vs Universal Acoustics. Molecular Echoeing vs Sound. Quantum Tunneling. Universal Enthusiasm … .
* The Paramount Law of Transformation. B;ological Blueprint of the Universe. Spac3 Program Since 1452. Quantum Watch. En3rgy; all we were, all we are, all we will be … .
* The Paramount Law of Transformation. Milky Way Design.

Dedication

*My work is dedicated to the disadvantaged people on our Planet,
with regard to education, equal rights, race, ethnicity, aspiration to perform,
as well as to my children ... ..*

*Book is dedicated to the people, who share fascination of the Universe,
Human, and Intelligent Design, Liberties, as well as those,
who were next to me, when I needed (AO).*

Vicky, Jacob and Jonah, I thank God for you everyday ... .

Marek „Mark" J, Wagner

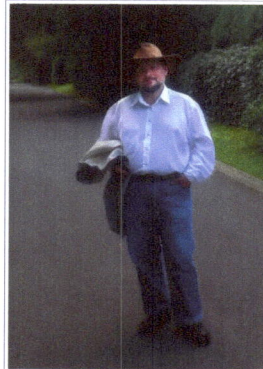

The Paramount Law of Transformation ($\tilde{\omega}$)
Magnificent Awareness

The Paramount Law of Transformation, Law of Everything can be defined as ever existent molecular (energy) motion, interactions. Molecular motion itself activates laws of physics such as gravity, nuclear fusions, diversity in density of particles.

Time is non-existent. Sequence of The Paramount Law of Transformation represent accurately space and man within (as well as everything within).

The notion of so called empty space, vacuum is false. Space, de facto, is the matter. Every square inch of space is filled with particles more or less dense, yet, never empty.

The beauty of *The Paramount Law of Transformation* is manifested in varieties of density of the matter as well as energies.

The *Paramount Law of Transformation*, the sequence, can be accessed from a different directions and stages (sequential imprints in space). For example man can access historical, past sequence as well as future (so called „time" travel).

Man can and will travel to the distant stages of sequence, historical sequential blocks.

Sequence is the past and the future. Present does not exist, de facto, due to the fact, that sequence is in constant motion.

Sequence doesn't end, sequence is infinite, yet, sequence transform into the progression of sequential pattern, blocks or sequence through the transformation stage becomes a new pattern, new sequential entity.

The shape of sequence is resembling spiral (twirl if you will) identical to the DNA spiral (space, motion, particles).

Human DNA precisely illustrate *The Paramount Law of Transformation* as well as *Magnificent Awareness,* logical sequential information, DNA pattern of purpose of existence.

Computing represent another example of *The Paramount Law of Transformation* as the progression of pattern, where is embedded space, motion, molecules, energies, laws of physics including gravity, electromagnetism, electromagnetic field.

Human vision precisely illustrate ***The Paramount Law of Transformation***. Molecules, lumens are being transformed into the electrical current and subsequently form 3D projection.

Human brain is absorbing molecular particles which are being absorbed, transformed and are projected back.

Human body function according to The Paramount Law of Transformation (electricity, magnetic field, transformation of molecules) according to the sequential pattern, which has nothing to do with so called time, but the sequence of light, which varies throughout the Universe, as we know it.

Equation of The Paramount Law of Transformation

$$\tilde{\omega} = s(m + n)n$$

(space, matter and motion (∞))

$\tilde{\omega}$ -The Paramount Law of Transformation (Ŏ- information, the smallest molecule in the universe as we know it). Logical information is forming the world. Information is complete, yet, is evolving into another form of complete stage within the Paramount Law of Transformation. Information, encoding of the reality, de facto, sequence.

s - space

m - matter (molecules)

n – motion (*3*)

     Important: information is the propeller and essence of the Paramount Law of Transformation, yet, information is already embedded in each molecule as well as within as well as in motion of space.

Space

Matter   Motion

(triangular stone was found in the mountains in 2014)

Light

Light

Light is common reality, phenomenon observed by senses as well as imagination. Modern science is describing and defining light, yet, light still holds so much mystery and unanswered questions.

1.***How light, a particle without mass, is generating energy to travel for billions of years.*** Where at the same time another source of light, a bulb, is generating similar effect, yet, disappears within seconds. Does light properties vary by the source of its projection ? Light is the source of information which interacts with its opposite so called „dark matter". The segmentation of the universe, as we know it, provide basis for such simple, yet, brilliant system where particles of light generated by powerful event, create wave of transformation of data between light and dark matter, and in this instance, wave progression can reach Earth even from distinctly remote locations. Shadow surrounded by light behind solid objects in space, provide another clue about the subject.

2. ***What is the Paramount Function of Light ?*** Does light have special quality in the universe as we know it, and how we can define this quality.

Light transforms universe as we know it and its influence is cyclical. Light is not present all the time because universe as we know it is defined by the compatibility of opposites, in this instance light vs darkness where darkness manifests its mirror image in opposite to the properties of light, yet compatible with each other.

The Law of Compatible Opposites is, de facto, fundamental in universe as we know it, yet, opposites are designed as compatible to each other.

By compatibility of opposites, in this instance light vs darkness, particle of light, lumens transform its opposite, darkness that is. This phenomenon, predictable chain reaction, light and dark particles are the source of motion in the universe, where dark matter transforms into the light (some kind of fuel). The compatibility of opposites is also observable in human being. Human being is the precise model of the universe including ***The Compatible Law of Opposites.*** Human is procreated from a tiny source of information as well as compatibility of opposites, from this tiny source of information emerges a human being, through molecular progression of cells, vast variety of specialized cells, where at some point this process is replicated.

Light vs darkness. ***The Law of Opposites*** is also observable as the model of psychological interactions, but this is another matter. In fact, universe as we know it, is defined by ***The Paramount Law of Transformation*** which is implemented in physical aware as well as unaware existence.

3. ***How to define light, particles without mass, yet, profoundly projecting physical qualities, illumination, in this instance.*** We can not dissect light phenomenon by only physical, molecular qualities and properties. Light is the code of information which interacts with other particles (information), at the same time transforming it and is becoming the subject of this transformation. Light is the source of information, as any living aware and unaware manifestation of existence. Light as the source of information has no beginning and no end, because is morphing and transforming from one phenomenon, data, into another, yet, embedded in the sequence, sequence which is predesigned to replicate in progression of mutations.

4. ***How Light Sculpt Reality.*** Reality is what we perceive as reality, yet, visible light is just the spectrum of the entire phenomenon which is hidden from human senses, unless explored by specialized equipment. And here we can use human model to illustrate it: lumen (the source of information) enters human eye (specialized equipment), subsequently light is transformed into the electricity (the same source of information, yet, transformed to penetrate human brain, and subsequently creates projection. Light is the source of information, de facto light does sculpt reality by transforming itself into another form of particles.

Sequence: light + eye + electrons + algorithm = precise projection of reflected light.

Yet, this is just the obvious, de facto, light (the source of information) is absorbed by the entire human body, not only eyes, and transformed into the other essential functions without which man would not function. Same principles apply to all forms of life or at least most complex.

5. ***What is the Real Speed of light ?*** Is the speed of light constant ? Speed of light is not constant. Light as the source of information performs optimal to the environment in which it does function. One of the examples of speed of light, besides space, light as the source of information is laser, fiber optics, optical tubes, including sub atomic particle accelerators in USA and Switzerland/France. In the universe as we know and according to ***The Paramount Law of Transformation*** nothing is, de facto constant, except when certain physical phenomenon is defined by precise interaction or absence of interaction. Rock changes according to temperature, exposure to light and moisture. Rock is also a source of information, solid, dense matter, which does change and interact within and beyond of its physical structure.

6. ***Does Light Changes Direction ?*** Light can perform in a straight vector type line, light can bend and refract. Examples are light particles within sun's core as well as on the sun's surface (where certain particles are stuck inside for billions of years, where others speed away), fiber optics, optical tubes, particle accelerators.

Light defined by matter, becomes self defined phenomenon and subsequently transforms into other forms of compatibility.

7. ***How Long Light Particles Exist.*** It varies where light performs, for example space/vacuum, intensity of projection as well as the source of projection. Light projected in the universe performs accordingly, yet, if environment changes light changes as well.

Light as the source of information does not manifest constant speed but compatibility with the environment it performs. Germans argue that light exist for approximately 3 years reference time, which is 10_{18}. This is not entirely true. Light performs by the definition of the intensity (force) of projection and the source of projection as well as density of matter. Light projected from a light bulb disappears within seconds, bright and powerful light projected by thunders also disappears within seconds, light projected from nuclear projection disappears within minutes, light projected by nuclear fusion within sun's core still shines after billions of years. Why ? Because light doesn't perform only as the particle but the source of segmented information which transforms the opposite segment of compatible source of information „dark matter", which is the opposite, yet, compatible (fuel) with regard to light.

This is essential part of ***The Paramount Law of Transformation***, compatibility in opposites, at the same time gracefully implemented in human design (physiology and procreation).

8. ***How Many Lumens Are in One Square Inch in Space?*** It depends on factual projection of space light ripples, yet, square inch measurement provides important data with regard to intensity of projected data called as light.

9. ***Light vs Human Body.*** Human body is changing and transforming (biological particles), where most of the cells throughout life are being replaced by new ones. Similar process is observable in the universe, including light. Light, data (same as DNA) transforms other forms of data, opposite data and at the same time compatible with the opposite.

10. ***Light and Absorbtion vs Compatibility of Opposites.*** We know that some materials efficiently absorb light, which disappears, yet, in universe as we know it, light is absorbed with the certain degree of intensity of projection, yet, light is designed to interact and transform compatible opposite, dark matter for example. If universe as we know it would absorb light as some materials do, than universe would have function according to the different model or perhaps would not exist. Human as well as Universe function by implementation of brilliant idea, motion and transformation of compatible opposites.

The Law of Compatible Opposites within ***The Paramount Law of Transformation.***
The Law of Compatible opposites is de facto essential propeller of cellular and atomic particles through progression of procreation where progression of procreation is manifested in aware as well as unaware existence.

11. ***Light vs Segmentation of Universe.*** Compatible opposites interact with each other in a variety of ways, as segments of data of opposite and compatible opposites.

12. ***Energy of Light Equation***

Mr. Planck's equation $E = hc / \lambda$
E - energy of photons.
h - constant (means absolute and can not be broken naturally, hmm).
c - speed of light.
λ - wavelength of light.

We have proved already that speed of light is not constant, than above equation in not complete and accurate. Mr. Planck's equation can be applied to certain situations, yet, his equation is not, in my opinion, compatible with the notion of ***Universal Properties of Light***. As of today, I'm still working on this equation, equation which will reflect true nature and properties of light as well as of potential, including speed, direction.

The Law of Compatible Molecular Opposites and Energies

Light vs Shadow vs Darkness.

To understand light we need to study properties of shadows.

1. ***Light vs Shadow.*** Shadow is formed by the process of progression of light sequence as well as reflection of light from various surfaces. Shadow is also manifesting segmentation system in the Universe as we know it. If room is illuminated by light, in various predictable locations will be formed shadow, which emerges by interaction between light and dark matter. Solid objects prevent light from interaction with dark particles. We can also observe degrees of shadows as well as degree of illumination, which clearly illustrate exchange of data between light and dark particles in segmented molecular Universe as we know it. De facto, light transforms periodically properties of dark particles. The Compatible Law of Opposites allows to interact between light and dark matter in space densely filled with opposites such as light and darkness. If light is without a mass than dark "matter" is also without mass because it's compatible in opposition with light so with great degree of possibility represent similar properties as light, yet, opposite and compatible.

2. ***Formation of Shadow Data.*** Shadow is formed by lack of interaction between light and the opposite, compatible and at the same time opposite matter.

3. ***Degree of Shadows***
       vs   = indicate ***degree of interaction*** between particles of light vs dark,
   Degree of Light   as well as transformation sequence.

4. ***Physical World vs Human Being as The Model of the Universe as We Know it.*** Physical world reflects identical properties or at least is showing similar dynamics, after all, human being is a collection of blended living molecular elements. The notion of compatible opposites is also observable in human emotional state as well as spiritual, psychological (personal, social, global). The very same principle defines, with comparable degree of similarity, not only physical interaction in molecular world but social interactions as well. Fundamental difference is free will which defines human interactions and here similarity between human being and physical world no longer applies. Yet, this is the notion for a new chapter. Light, in spiritual and psychological scheme defines also a state of awareness, which translates as the sequence toward perfection and the observable status quo of this process, which varies from person to person. Free will defines the state of spiritual, emotional, ethical illumination in social interactions. Based on The Law of Compatible Opposites we shall apply same kinetic interactions with regard to molecular, living, physical substance, yet, not aware, versus kinetics of living molecular existence, aware, because above law is universal throughout the universe as we know it.

5. ***Motion of Light vs Darkness.*** It seems that the fundamental difference between light and darkness in the Universe as we know it is this: illumination (light) is changing its position while dark matter does it with only marginal degree. Dark matter is not motionless but much less "mobile", that's why the entire Universe as we know it is filled with dark matter and that's why we can come to the conclusion that there are boundaries of the Universe. Perhaps the boundaries of the Universe as we know it is just the absence of interaction between dark matter and the light.

6. ***Where is the end of the Universe ?*** Difficult question because light visible spectrum to the human eye is just a fraction of interactions between light and dark matter, yet, when those interactions are absent there is the end of the Universe as we know it. Light is essential quality in terms of existence, molecular, biological.

7. ***Boundaries of the Universe as we know it.*** If light vs darkness sequence of data is no longer maintained than the entire model of the Universe is no longer supported and there are the ***Boundaries of the Universe*** as we know it.

8. ***Does the Light Travel ?*** It seems, according to present frame of awareness that light does not travel, per say, but interact in an exchange of data between photons and dark matter. Photons transform sequentially dark matter. This process is also shading light on notion of shadows, its properties, degree of interactions, lifespan.

9. ***Universe vs Infinity.*** Infinity in true sense is a process of procreation of transformation of data within molecular world including light vs dark matter, as we call it, illustrate this process perfectly.

10. ***Light vs Degree of Illumination.*** By observing Degrees of Light we can also assume that the sequence of so called other worlds is based on scheme of sequence light vs dark matter where light progressively dominates over dark matter and is "competing" with degree of illumination until the sequence of illumination is completed. Above conclusion can be easily illustrated: life, molecular, organic and aware is progressively complex and beautiful with parallel progression of light, we can then assume that the same progression toward perfection applies in Degrees of Light beyond Universe as we know it.

Above is a draft which may change over time, yet, you are invited for the discussion.

     Sincerely,

Marek „Mark" J. Wagner

The Paramount Law of Transformation

vs

Speed

The Paramount Law of Transformation vs Speed

Does light represent measurable property ? In certain environment yes, yet, speed is much more than reference of light vs space vs sequence, speed in molecular sequence represent "an instant". An Instant Speed is the true potential and physical property of speed. An instant Speed is Human destination without physical restrictions.

An Instant Speed is the speed to reach any destination in an instant including awareness which is also molecular. This is a true nature and a real possibility of speed, human destiny.

Human perception is bound to the past. Every modern equipment on Earth and in Space is performing in relation to the past sequence. We are unable, as of today, to witness real-time phenomenons. Hubble telescope is penetrating past, yet, future is beyond reach.

An Instant Speed is the solution for this problem, and the solution is within human brain and the wavelength of thought, the fastest particle in the universe as we know it, derivative particle of light. Light is just the beginning toward the greatest human adventure, An Instant Speed ... .

The Paramount Law of Transformation vs Biological Model of the Universe

The Paramount Law of Transformation is evolving still. Biological Model of the Universe, Light vs Darkness, Theory of Shadows, Black Holes, Law of Everything Equation, Instant Speed and the Collection of Laws such as Law of Compatible Opposites, Symmetry and Anti Symmetry, Geometry vs Anti Geometry, Aerodynamics of Transformations, the Law of Beauty and Aesthetic Existence defined by light which eloquently is verbalized in everyday life as well as on Hubble Imagery and few others.

The Paramount Law of Transformation does, de facto, reflects the Universe as we know it, and the man as the essence of all laws in nature, de facto, a model of the universe, simplicity in complexity. Few simple laws of the universe define all existence, yet, if someone would tell you: objection :-) , please repeat after me, whisper to the Divine ear what you've just learned, and you'll see what happens, next universe will be just as I have told you ... .

Yet, the ownership of science dwells within the force which is humble and grand, beyond human perception, we are just on the path to rediscover science which already has been invented, long time ago, including light, transformation and motion ... .

Mirror image is a faithful reflection of its own reality, compatible, yet, opposite.

Universe vs Expansion

Universe, based on biological model simulation as well as logic is not expanding but works through expansion and contraction. Perhaps right now is the period of expansion and subsequently Universe as we know it, will contract. The model of expansion and contraction is in place. In addition Universe is circling around its axis. The Paramount Law of Transformation tolerate nearly anything with one exception, static state.

Supernova, violent expansion and equally dynamic collapse, in some instances Vortex of Matter, Black Holes indicate a phenomenon of expansion and contraction same as the creation of the Universe but in much smaller scale. Perhaps. At the final stage entire Universe will collapse to emerge as the new reality ... .

The Paramount Law of Transformation does not tolerate static state, unless to briefly cool off molecular emotions (state of sleep), yet, always in motion to evolve again and again through brilliant recycling of molecules, matter, energy. Can we say, every man is like a different galaxy where all elements of the universe are present, manifested ? Yes, certainly ... .

Universe vs Vortex

On Earth vortex is created through interaction between cold and hot air. It seems that this force is so powerful that nothing can escape from its path.

We can than formulate notion that vortex in general is created by interaction between compatible opposites. Compatible opposites, common in Universe as we know it, is light vs dark energy.

Why than Vortex of Matter is present in one location and not present in the other. Collapsing stars, where density is so great that in some instances exceed thousands of Suns within, yet, the size is smaller than sun's indicate that expansion and contraction of matter plays a key role, same as human heart. Do the center of the universe is the "heart" in terms of biological model of the universe, where light and dark matter and radio waves (already preprogrammed and defined in terms of properties) is the "intelligent mind" of the Universe, because all particles are, de facto, information either molecular with mass or waves of energy).

Universe vs Creation

How was Universe formulated. Universe was cold and static until powerful force of the creation put in motion entire blueprint of dense matter. This process had its initiation from the center of the universe. Subsequently expansion was initiated where gravity hold everything in place: hot light vs cold light (dark matter), in other words, opposite information spin around in fantastic vortex of life spin where man has to find its place and destiny.

Universe is filled with matter, in every square inch, yet, the density and quality of matter define its dynamics and performance, according to the Biological Model of the Universe. Density of matter is parallel with kinetics, the denser matter becomes, the more dynamic properties and projection of energies.

The Paramount Law of Transformation vs Energy

Auroras display approximation of molecular, energetic interactions in space. Behavioural shift in colors, shapes, liquid, harmonic kinetics also reveal invisible, yet, ever existent energies into visible physical world, and apparently this is just a fraction of the dynamic nature in Universe as we know it.

Auroras projeciton on North and South Pole also reveal waves of energies, energies which define physical world, shape and variety of interactions, simple, yet complex.

In addition, energies and interactions are not only exist and essential, but display aesthetic qualities, qualities which form, profoundly, human perception, perception, which is seeking beauty in nearly every moment of human existence.

Space vs Energy of Hot Light vs Cold Light *(Dark Energy)*

What would be the ratio between hot light and cold light (dark energy). Interactions of energies suggest (Aurora for example) that this relationship is not constant, yet, varies within certian parameters, which hold everything in place. This realtionship define Universe, universe which is not excessivelly hot or too cold, as of today. And again, the relationship between compatible opposities is not only existent but fundamental in Universe.

Universe vs Space and Time

The property of time is not essential for existence of space. The essential quality for space, molecular, filled with energies existence aware and unaware, is sequence of interations, transformations.

If we would ask the Universe, time, what's the usefull application. Universe would certianly bounced back with an answear: „universe doesn't care about time but sequence". Yet, time is a common, plesant accesory in our existence and awareness, as of today.

Everyday we hear from all directions: „timeless universe". Even if Universe is not timeless the frame of existence is irrelevant.

The Paramount Law of Transformation vs Compatible Opposites

The foundation of The Paramount Law of Transformation is the property of the compatible opposites. That's the defining element as well as moment of sequence in space and sequence of existence, molecular aware or molecular unaware existence.

Please do not use a term timespace because is misleading perception and scientific awareness. The true reality is sequence within space. So called time is just an accessory which you can wear like an elegant jewelry, yet, it's not essential, if not harmful, in terms of proper alignment of understanding. Sequence within space shines much more eloquently like exquisite jewelry, than so called timespace.

Progression of Matter (simplified).

- Simple Matter (high energy / high capacity information)
- Complex Matter (evolving into the complex sequence and energies)
- Supercomplex Elements (formation of galaxies)
- Intelligent Matter
- Biological Matter (complex progression into the blend of particles within sequence)
- Intelligent Biological Matter and Awareness
- Artificial Intelligence (rapid acceleration of artificial intelligence through progression and replication of data and subsequent progression beyond reach for homo sapiens
- Possible end of human race, as we know it, into the biologtical, intelligent robots.

Sequential Pattern within the Universe.
DNA of the Universe. Expansion and Collapse.

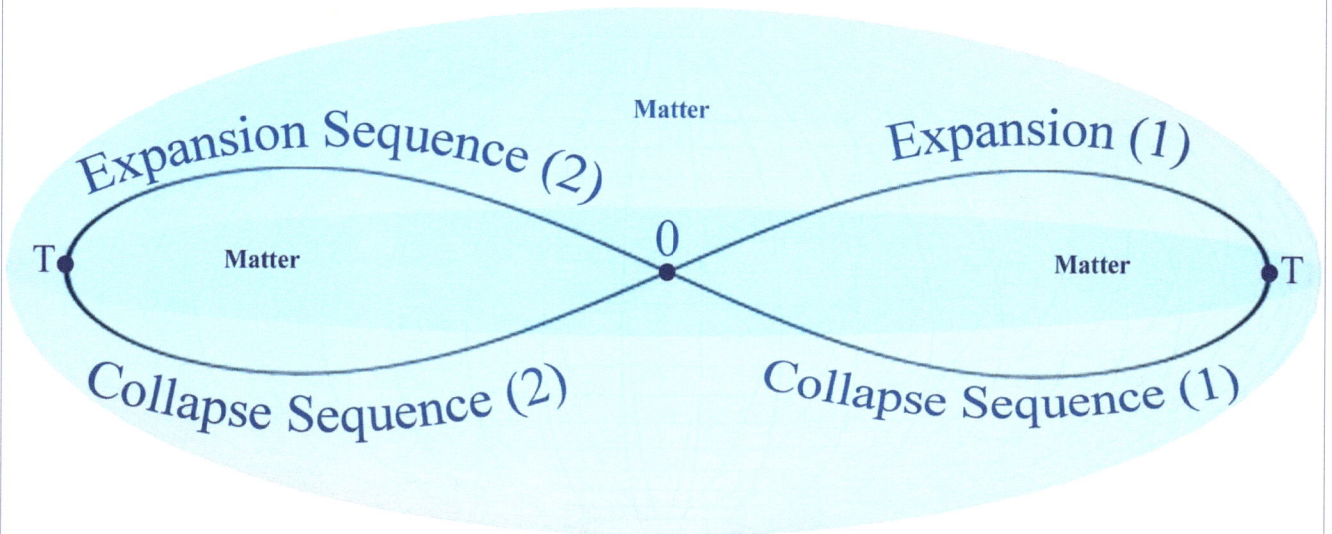

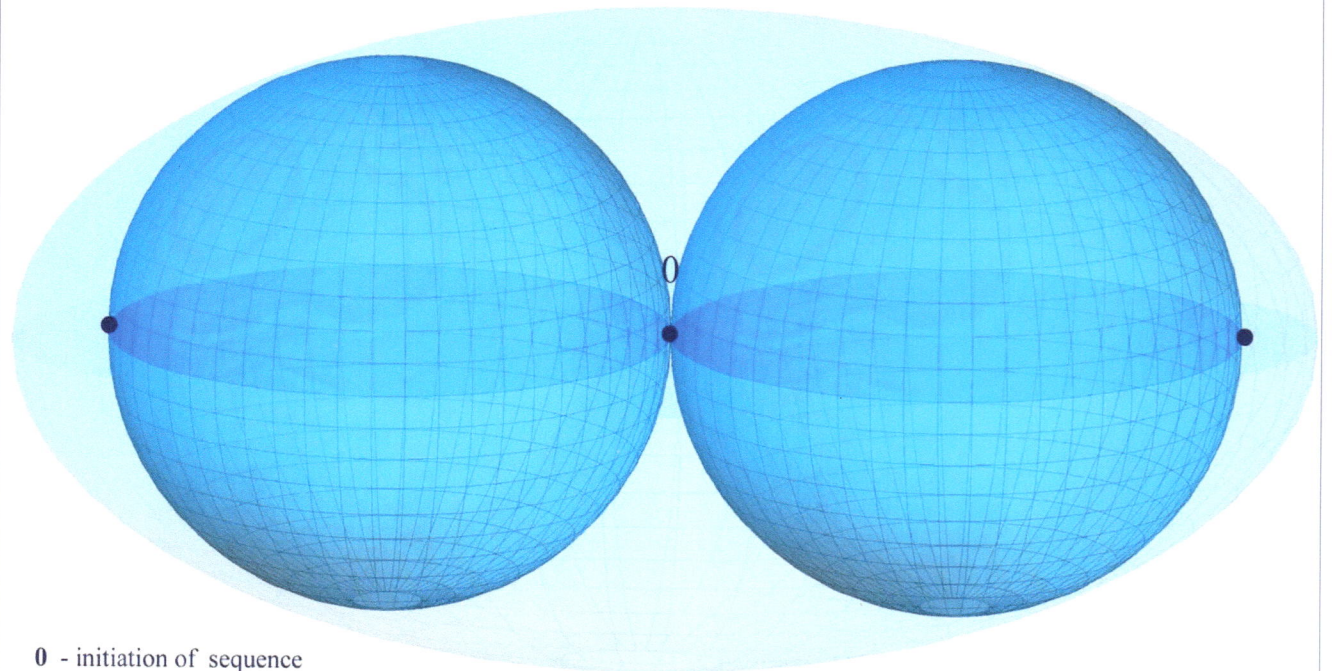

Expansion Sequence (2)

Matter

Expansion (1)

T

Matter

0

Matter

T

Collapse Sequence (2)

Collapse Sequence (1)

0 - initiation of sequence
T - the tipping point

**The Paramount Law of Transformation: Biological System of the Universe.
Multidimentional. Multiverse Existence.**

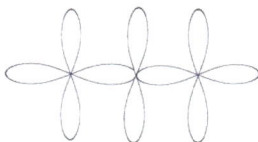

Fragmentation of the Miltidimentional Worlds - Multiverse
(based on density od matter and energies - possibility).

The primary function of so called "black holes", is distribution of matter. It's a very complex transitional recycling system of particles, galactic gear box. Distribution of matter is the essence of Paramount Law of Transformation.

Time doesn't exist, instead, there is rythm of sequence.

The Paramount Law of Transformation reflect The Law of Imperfection, which is perfection within molecular structure of the world, cosmos, at the same time evolving to the higher level of complexity ... .

The Paramount Law of Transformation.
Molecular Communication.

As I have already indicated, moleculs are, de facto, specific information, data, which contain certain properties. One of the basic function of data is communication. Molecules communicate with each other. Communication goes within and beyond molecular environment. Subsequently, in many instances, communication is evolving, where unification of the molecular world is impossible to ignore or disregard.

One of the classic examples of molecular unification is the Solar System. Subsequently Milky Way is in motion as well, circling, twisting gracefully along with gravitational force of stars, molecular vortex (so called Black Hole).

The entire Universe is evolving through countless Whirlpools, vortex sequence. Universe is a complex, yetm simple system of Whirlpools, which communicate between smaller and larger communities, galaxies, until the unifying energy will be completed or will reach a certain point, a tipping point, as illustrate chart below.

It seems that information, data, is never lost, yet, is going through rigorous and fertile system of transformation, stages, which define existence within Universe as we know it and beyond.

The Paramount Law of Transformation

Gravity

Gravitational Forces.

What we know:
•space
•matter
•motion
•sequence of clusters (fragmentation of Universe as we know it)
•light (including Law of Opposites)
•cold light (dark matter)
•density of matter

•*Gravity Social Properties of Universe (Matter and Energies)*

We do know, so far, that light and cold light represent paramount equivalency in terms of existence within the Universe, molecular existence, unaware and aware, including biological, that is, based on the Law of Opposites.

I tested both experiments before the first grade. When I got my first bicycle, I used nearby hill in small village of Szczytniki to learn how to ride a bicycle, apparently, gravity did all the hard work for me. Somehow, via intuition, I understood the benefit of gravity.
In 1969, once a week, my mother used to send me to the local farmer to bring fresh milk. Every time, while on my way back home, I performed experiment as the boy on attached illustration. Somehow, by intuitive engagement, I understood that milk must remain inside a small aluminum jug, even when it was upside down. Yet, at one time, milk spilled nearby pastures, because I was experimenting with circular speed.
We have here not only simplified illustration of gravity, but acceptable ratio, which holds everything together.

Motion

Why motion exists ? Because Universe, as we know it, is in constant motion (molecular, clustered) and the whole Universe, as a whole, in motion as well through expansion including circular motion). Motion is embedded in every molecule and energy in Universe, Social Molecular Universe. Motion is the DNA fabric of living Universe.

Artificial Weight in Space

Artificial weight in space can be achieved through motion, yet, motion in Universe is the result of interaction between light and dark matter as well as molecular densities.

Gravity

Gravity (de facto, information embedded in energy of light, cold light, density of molecules) is the result of interactions between light and cold light, as well as molecular interactions in space, filled with matter, which varies in density.

Gravity represent simple, yet brilliantly eloquent ratio between motion in space initiated by light and cold light, as well as density of matter = cluster of matter and energies, gravity.
With regard to gravity, we observe fragmentation of gravity in space, from weak to strong to extremely powerful forces of gravity (vortex of matter).

Galaxies are, de facto, clusters of gravity associated with variety of matter.
Gravity is defined by the precise interaction between light and cold light in the universe, as well as coefficient interaction between density of matter within vicinity. When interaction between light and cold light is no longer detectable, including at the molecular level, Universe no longer exists as the living matter and subsequently the gravity (universe "freezes" in all its fantastic possibilities and incarnations).
Structure of space is defined by interaction of opposites, yet, compatible clusters of information, in this instance, light versus cold light. Without interaction between energy of light, and the opposite energy of dark matter (spectrums of energies of light), gravity is non-existent.

Gravity

Gravity is initiated at the precise moment, when ratio of interaction between light and dark matter, as well as molecular density of available elements within vicinity, reaches the tipping point, which define motion of matter, spiral motion, de facto, toward inner core, which contain concentration of heavy metals.

Gravity is indicating compatible interaction within Universal blueprint of matter, including variety of elements from heavy to ultra light energy of thought for example, which is not bound by gravity, yet, does function within.

Simplified Fragmentation of Gravity in Space

- Space: weightlessness (gravity force varies)
- Planets, Stars: gravity
- Galaxies: gravity forces are pulling matter inward, holding molecules and energies together (***Social Property of Matter and Energies in the Universe***)
- Vortex of Matter (black hole): spiral motion and transition from one cluster of information into another.
- Boundaries of Universe: the weakest gravitational forces, yet, existent, which will enable Universe to contract (collapse) after expansion (after raching the tipping point of expansion).

Searching for Life in the Universe as We Know It

To search for biological life, similar to our own, I would propose to define certain criteria:

- Biological living forms of life (from extreme minus temperature up the the extreme hottest)
- Speed of planet Earth (inner core, speed around of its axis, orbital speed), based of elemental frequencies (liquid formulation of metals)
- Distance from a nearby star: searching for approximation with regard to the distance between planet Earth and the Sun as well as the Moon
- Graviational approximation of equivalency between planets, based on Solar System within the Milky Way Galaxy
- Earth Talk: projection of frequencies of electromagnetic as well as gravitational waves.

Above represents the ***Formula of Existence***, approximation, where biological forms dwell within the Universe, as we know it, yet, not a defintion of other, possible, forms of existence.

Gravity Equation

Gravity: Motion, Speed, Mass (Density), Space, Distance.

$$G\tilde{\omega} =$$

$G\tilde{\omega}$- gravity (variable /non constant throughout the Universe)

$\tilde{\omega}$ -The Paramount Law of Transformation

$\breve{O}$- information

s - space

m - matter (molecules)

n – motion (*3*)

p - speed

VORTEX OF MATTER

Vortex of Matter (Black Holes)
Water Incarnations
Biological System of the Universe

Water, besides air, is the most common type of energy, essential, de facto. Water falls, as well as human eyes illustrate the dynamics of Vortex of Matter in Universe, as we know it. Transition of matter from one form to another can be illustrated on biological model of the Universe, including human biology, where energy of light is entering human eye and is transformed.

The dynamics of the transition process illustrate water falls, where assembly as well as disassembly of particles is happening in front of our eyes.

Water is the only known natural substance on Earth, where three physical states occur, such as solid, liquid and gas. Ice, water, fog, steam, clouds, dew, volcanic steam, salt water, fresh water, de facto, one substance is the subject of variety of incarnations.

Water falls, human vision illustrate how our Universe perform beyond the rim of known.

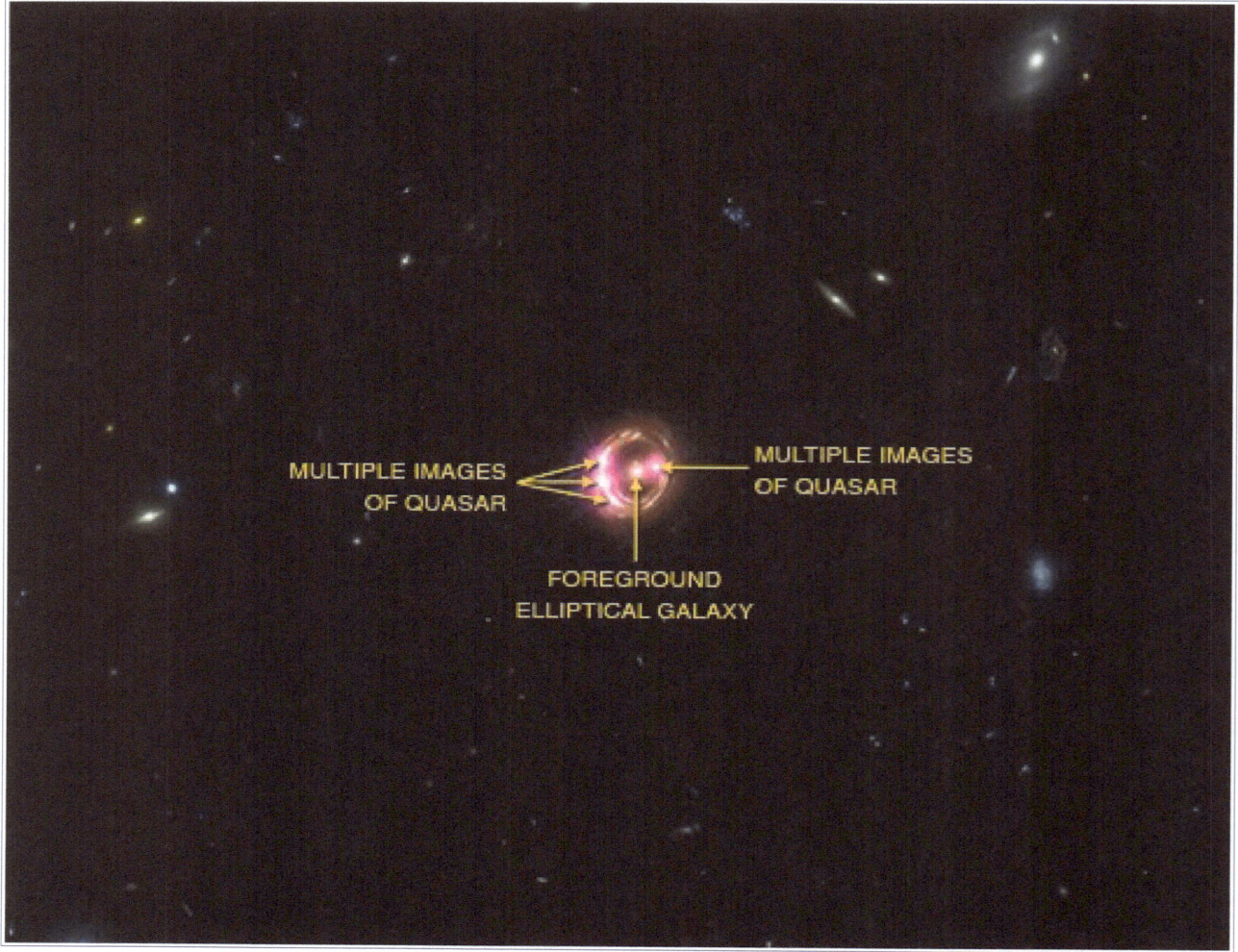

MULTIPLE IMAGES
OF QUASAR

MULTIPLE IMAGES
OF QUASAR

FOREGROUND
ELLIPTICAL GALAXY

Notes and Thoughts

Natural Patterns

*   Philosophy can perform without science, yet, science can not exist without philosophy. Both phenomenons are essential in terms of proper alignment of awareness …

*   "Nothing" can exclusively refer to space, where light is not present. This is my definition of "nothing". "Nothing" is non existent …

•*   The most powerful force in the Universe, as we know it, is **Potential,** the most fertile science ever existed … . I would like to introduce **Universal Law of Potential** …

*   If beautiful theory is non provable, perhaps it's time to make it real, because man, by the grace of God, will create new worlds, and new awareness … (as long as is ethical)

*   I wonder, does Universe shape human awareness or perhaps human awareness is shaping the UNiverse, as we know it. Above is the essence of all questions …
Shape of Universe vs Frame of Awareness, the essence of all questions …

*   Light is the most precise and beautiful tool to measure Universal awareness …

*   Its seems, that everything we know, to date, is journey through spectrums of light, visible and invisible, within and beyond man, Universe, as we know it, since the beginning of sequence, and even prior to beginning …

*   Does density of light (spectrum) define existence ? Certainly. If Universe, as we know it, was created, formed nearly 13.82 billions years ago, and subsequently Earth was miraculously formed approximately 8 billions years later, than we can come to the conclusion that the development of matter, until it reached biological state, is the most advanced form of matter within the Universe …

*   The beginning of sequence of the creation ("time" of creation) indicate evolution in space until it reached biological state …

*   What is the background of Universe, as we know it, background beyond the smallest particles ? Progression of patterns, computation of forms, self replication … .
       If you look very closely, deep inside, you will see "noise" of information, which is no longer seamless, but defined by regular, geometrical pattern (beyond circles), kind of spider web. Yet, the question is, how spider web is made of ?
Above indicate that this is virtual world or structured on molecular geometric principles, or both …

*   Does spectrum of light, the opposite, yet compatible "cold light" is the background of the Universe, on which hot light performs its pirouettes, and dances until exhaustion of senses, to return to the point "0" , the new beginning of the new awareness ? Yes …

*   Perhaps we will not grasp the essence of life, until we come up with precise equation of an **IDEA** … . The notion of an Idea is present in DNA progressive computation, yet, how miracle of an Idea became a reality, intelligent sequence embedded into the matter is fundamental to explore ...

*   I believe in light, spectrums, which shape and miraculously carve space. Light is the carrier of life, with embedded information about human DNA. Divine Universe was destined to carry life from Day 1. First was word, awareness, and Fiat Lux, Let it be Light …

*   Modern science, scientists, are the Apostles of progress, as long as the Divine whisper of Universal Potential, moral spine, is not disregarded in notion of intellectual liberties ...

*   ***Human, Blueprint of the Universe,*** Biological System also represent startling similarities, when it comes to transition chapter. Physicians, as well as physicists describe very similar phenomenons with reagrd to Vortex of Matter (Black Hole), as well as transition of so called life after life.

Human brain represents nearly identical scheme as physical world of collapsing stars. In addition, if we will be able to explain entire physical transition of collapsing star (Vortex of Matter), we will be able to tell, what happens with human awareness after transition is complete.

American studies of military pilots in 1960's also describe that human brain, at certain speed is projecting similar, if not identical, "tunnel syndrome" which is, de facto, a vortex within biological transition moment.

Human brain is going through identical or very similar changes as Stars, and subsequently Vortex of Matter. In addition human is experiencing kind of "explosion of energy during transition", where all memories are compacted in an instant, same in sequence of Vortex of Matter.

Human is a precise map of the Universe, as we know it, where biological existence is built and blended from the same particles as Universe, including projected laws defined by ***The Paramount Low of Transformation.***
Above can not be challenged, in my opinion, because human senses record similar data, as sophisticated instruments in space and Earth's orbits (transition phenomenon: human brain vs transition of Stars in space).
Now I've got to find traces of so called cold light within, or the compatible opposite which define the compatible opposite … .

Friends, do not seek no further, than within self ...
MJW

Quantum Mechanics and Other Affairs ...

Last summer (2014), I submitted an image, where the same person appears twice. This isn't exactly quantum physics, where electron can appear in two locations simultaneously, yet, quite original and well crafted to use as an illustration for quantum physics. Yet, my intention is to use this example as an illustration of Quantum Sequence, including possibilities and potential, in terms of physical accessibility.

360 degrees image, where the same man appears twice is proving that sequence allows to access at any location, at any chosen fragmented sequence. Frankly, attached image is about sequence, than time, which is irrelevant for Universe, yet, it could be associated with well designed and fashionable time piece.

In terms of quantum physics, projection of physical world, refers to awareness as well, which, de facto, is proving fantastic logical efficiency. Let's analyse TV projection.
Human, by implementation of logic and modern technology is able to use available resources, such as electrons, for example, digital projection. The very same principles, are manifested throughout the Universe, as we know it, but on much larger scale.

Universe is the 360 degrees, 3D projection tube, we could say (very precisely). Every time we learn to watch the show (projection), we discover not only logic, sophistication, but beauty of harmony, compatible with our own senses, including aesthetic sense, profound manifestation of awareness.
It's like „something would care to show the best face of the Universe".

In addition, compatibility of projection in Universe, as well as awareness of perception, is proving, that we are an integral part of the system.

Yet, it's not as obvious as it seems. Many of us, for example, while travelling, are exposed to meals or food, which is not compatible with your taste. This happens, when you are not programmed (brain) to certain types of foods, spices, cooking styles, even smells, flavors.
The reason why this happens, is because you are not compatible, de facto, your aesthetic sense, in terms of cooking, to the certain areas, cultural, geographical.

Human (brain, nervous system, awareness, perception) is programmed identical as Universe is, in every molecular niche imaginable. Human and Universe are compatible in every sense, including, aesthetic sense.

Above represents another logical clue, that Universe can not be comprehended, intellectually grasped and efficiently projected in mind, without new science: ***Biological System of the Universe***.
All laws, all mechanisms within Universe, as we know it, more or less, are embedded into the human being functionality, including the most advanced, sense of beauty, emotions, aesthetic perception, and profoundly important, proportions.

Why Universe „prefers" logical (not awkward) projection of qualities, while proudly projecting beauty of aesthetic awareness, including proportions ... .

Because is compatible with intelligence, that's why. We all can agree that Universe, as we know it, is logical, beautiful, *Universe is Intelligent*, in all aspects of existence, aware and unaware.

At some point, in the past, I realized that hot light interacts with cold light, similarly as in above picture.

Pond is the Universe, filled with cold light, the opposite, yet, compatible. Wave of hot light interacts with cold light, and transforms, profoundly, space.

Space, before *Universal Initiation,* was present, yet, it was cold, dark, without a nervous system, *Universal Mind*, waiting, de facto, to be transformed into the great physical omelette, perfectly cooked for scientists, human perception, seeking Universal recipe, generation after generation. All essential ingredients require knowledge, and this science is light, de facto, *Universal Initiation of Existence*, aware and unaware through *Universal Mind*.

Universe, unaware, doesn't care about anything, but its existence, is programmed that way, yet, it progressed into the physical, even biological, intellectual awareness, which cares about its quality, appearance, elegance, including aesthetics sense. That's the defining moment to recognize, while looking through the telescope toward distant space, worlds, you are embedded, which are within your own being.
Unintelligent Universe wouldn't create intelligent worlds, yet, profoundly *Intelligent Universe* creates profoundly intelligent worlds, expressions.

Universe is speaking with a voice of Universal message: **You are in me, I am in You.** It's true, de facto ... .

Yet, *Universal Mind*, light, is cooking superb omelets on both sides of dimensions, even on all sides, all dimensions …

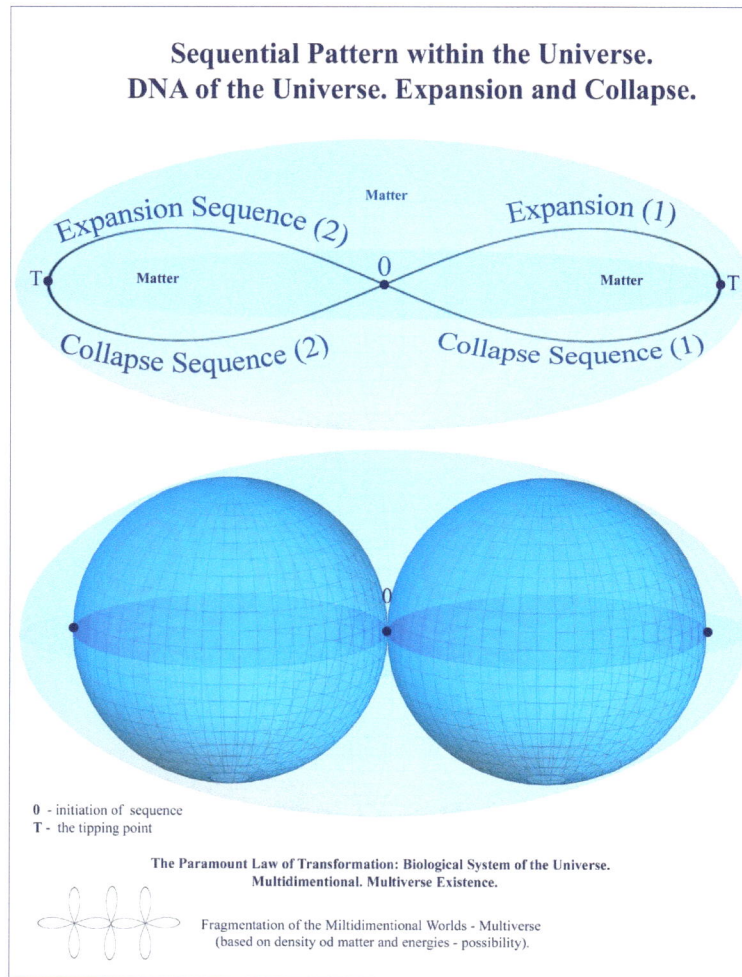

Sequential Pattern within the Universe.
DNA of the Universe. Expansion and Collapse.

Expansion Sequence (2)   Matter   Expansion (1)

T   Matter   0   Matter   T

Collapse Sequence (2)   Collapse Sequence (1)

0 - initiation of sequence
T - the tipping point

The Paramount Law of Transformation: Biological System of the Universe.
Multidimentional. Multiverse Existence.

Fragmentation of the Miltidimentional Worlds - Multiverse
(based on density od matter and energies - possibility).

Dimensions

1. Length
2. Heights
3. Depth
4. Sequence (Time)
5. Awareness of Senses (present in biological existence and a sequence toward Universal Mind)
6. Transition of world into worlds as in above illustration …
7. …

There are, it seems, 6 primary dimensions, which subsequently evolve, progress into the subsequent dimensions, compatible, de facto, yet, profoundly different in imprint of awareness …

Seventh dimension is Sunday. I tried all possible combinations, yet, beyond primary dimensions, I got replication of previous.

Dimensions also illustrate progression of awareness. I like the notion of frame of awareness, because is very much human, and profoundly sophisticated, navigating with an open sails of awareness, toward *Universal Mind.* Perhaps, some day, human will become frame of awareness, *Universal Mind*, in every sense.

We've got to follow *Universal Qualities of Space*, *Intelligent Universe* (because it's profoundly intelligent) in order to grasp the essence of Universe, as well as ourselves. It's not difficult, because Universe, physical, biological existence, is following certain patterns, I wrote about it decades ago.

Follow the *Universal Qualities of Space*, including *Intelligent Universe*, true, in every sense, and you will embark on journey with an open mind … .

Universe is Intelligent, de facto, isn't it ?

What do you see ? Universe is performing, profoundly intelligent, beautiful, logical, aesthetic, brilliant varieties of expressions. *I am in you, you are in me, you hear … .*

Why Man Exists … .

Definition of energy is an ability to transform. To me, this is the simplest and the most accurate definition of energy. Universe is based on simple principles.

Some scientists reffer to Universe, in terms of structure as "onion". Yet, within reach of awareness, including tangible senses, the model of the Universe, as we know it, is human being.

Man is the quint essence of the Universe. Within Human is embedded every solution, physical law, energy, blue print of Universe, molecular changes, including molecular regeneration of elements.

If we will understand self, human body, functions, molecular transitions, hierarchi of growth, miracle of Universe, „enzymes", that is, than we will have a model of the UNiverse, as we know it, and perhaps even more advanced understanding of existence, in front of our eyes. Virtually... .

Human, de facto, manifest, illustrate performing Universe, at the most sophisticated level, yet, the transition process doesn't end here, and we ought to find out subsequent, logical impersonations of matter including awareness … .

What is the future of Universe, as we know it. This question was produced by Mr. Hawking, whom I admire. Mr. Hawking, progression of Universe, as we know it, is showing important pattern, molecular progression, autmation, based on data, from "basics", to the most advanced in our reality, biological existence, including biological awareness of self vs molecular awareness (program/data).

All matter is a living matter, as long as is able to transform, the only difference is the presence of awareness. As of today, all matter is showing ability to transform, and apparently water is the most beautiful example, how matter, as well, as a source energy behave, transform. There is no such thing as the end of the Universe, but another path toward progression.

Molecules and energies are intelligent, if they progressed, up until the creation of Human Being, than the end of Universe, as we know it, is nothing more, than projecting progression toward another path defined by Universal logic.

If Universe is progressing, according to the above scheme, than the subsequent and anticipated progression is reaching the next, higher level of sophistication, until it reaches the tipping point, yet, when it comes to the progression of awareness, than ***possibilities are endless.*** After all, Universe is a simple, yet, demanding, in terms of self preservation through process self perfection, adaptability and progression to the higher level of sophistication, same as human being … .

Within human being is embedded entire Universe, and its history, including miraculous accessory of dialogue between fantastically blended and organized particles and energies, even at this virtual instant … .

It's possible that future is embedded as well within man, yet, we've got to learn how to read available data and interpret it. Human DNA could be helpful, in terms of predicting the future, or at least, it can provide some essential clues. There is no such thing as present, because squence is constantly progressing, yet, if we agree with this notion, than we ought to absorb another simple fact, the past is the future.

If the past is the future, than knowing what was in the past, we shall have an access to the future, as long as we will be able to be open minded … .

If we can't disprove any possible venue of progression, than it does have its place in discussion, until we will able able to define otherwise. *The Law of Possibilities* is simple, eloquent and inclussive.

If we can predict that within certain frame of sequence, we can perform certain task, than we can adapt this ability toward the future of human being and Universe, as we know it.

Biological aspect also illustrate and provide important clues through cellular progression of predesigned automation, cells (male and female), until sequence reaches levels of maturity (multilevel sequence within sequence): conception, birth, motion, speech, intelligence, replacement of cells etc.

All aspects of human existence, are compatible with Universal existence, molecular and energy, which shape miraculuously life, aware, as well as unaware: space, awareness, transitions, dramatic transformations, sequence of blending compatible opposites, intelligent data within particles, according to *The Paramount Law of Transformation* … .

Important question is as follow: *why human exist ?* Besides theological arguments, which can not be disproved as well, according to *The Law of Possibilities*, we can, once again, return to the same statement, as mentioned. UNiverse is progressing to the higher level of sophistication, and apparently Universe is intelligent enough to manifest profound sense of molecular harmony, ambition, attachment to symmetry, and equally important, attachment to the opposite, yet, compatible.

Does than UNiverse is expressing human qualities, or perhaps human is manifesting universal qualities, or both. Yes, both, because Universe was destined toward progression, same as man … .

I perceive Universe, as a living, intelligent, brilliantly progressing entity, made of intelligent particles and energies, which are aware abou its mission. Universe, which manifest unique quality, such as personality, based on molecular progression toward perfection, through progression of transformation of matter and energies … .
The facto, beginning of the Universe, as we know it, and the status quo, which can abosorb and be aware of, is proving just that.

Human biological existence, from initiation, until it reaches the tipping point, biological transition state is proving that the sequence of existence is progressing from limited, if not singular sources, (compatible opposites) to another, subsequent existence, which become a source of another, and another, and so on.

Please also analyze that we are related to singular ancestors, and it's a proven fact by available science. From singular sources, human society progressed to nearly 7 billion. This molecular, as well as biological pattern is indicating that UNiversal progression is infinitely fertile. The model of UNiversal progression is on front of our eyes, de facto… .

Human represent miraculously available data. Human is the reflection of the Creator. It is true, because human is the map of the Intelligent Universe. The notion of Human vs Universe, Universe vs Human represent interwoven, precisely blended parallels of existence, indeed… .

Memory

One of the common and fundamental quality in the Universe, as we know it, is the presence of the ***memory***. In every molecule or manifested form of energy is present data, memory of the design, the facto.

The paramount property of memory is existence. Every form of existence is bound by data, de facto, memory of the design. We can agree with notion that particles, manifestations of energies, are designed, no matter how progressed, at the same time we can not disregard or bypass the notion of intelligent design, because, as I have already indicated in previous comments, Universe is profoundly intelligent, due to the data, memory of design.

If memory is present, no matter how sophisticated, than the presence of intelligence, intelligent design, which blends physical world, including energies, molecules, is beyond any argument, no matter how design progressed from molecular into the more sophisticated progression of patterns. Memory (data) in order to occur, require space, energy, design progressed from fragmentation of the whole or as a whole in fragmented reality (logical).

If memory, ***data is telling, describing a story, organized in logical sequence***, and showing its blueprint, as well as function (purpose) in space, including motion and predesigned possibilities, than intelligence behind it is a matter of awareness. The Law of Possibilities ... .

In Universe, memory is commonly fragmented, due to the sophistication. Fragmentation of data, memory represent the quality of intelligent design toward simplicity and efficiency.

Memory reflect, profoundly, intelligent design, same as vast varieties of all data available throughout the universe, according to The Paramount Law of Transformation ... .

...

Progression of Intelligent Design and Existence

Space is filled with matter, energies, no matter how small. Once space becomes a reality, is parallel with existence of variety of densities, yet, some molecular manifestations wait to be discovered as well as energies … .

Origin of water, exciting journey to find sequence in space, where water became miraculous reality, is fundamental in search for universal answers. Water is older than Sun, which supports biological life, including aware existence. Water, according to modern estimates, is nearly 5 billion years old.

Glass of water, represent an example, de facto, of Intelligent Design in the Universe, as we know it. If water is older than the Solar System, that we have a strong indication that Universe was destined, through transformation, according to *The Paramount Law of Transformation*, toward biological existence. If we will be able to answer the question, how old is water in Universal terms, we will know, perhaps, how advanced are similar forms of life vs human civilization. The domino effect of experience, and knowledge in space, with regard to Earth, can serve as a model to intelligently verbalize possibility of scientific progress of possible civilizations in Universe. If.

Biological existence in space indicate Intelligent Design, as well as strong and scientifically feasible possibility, that Earth is not the only Fertile Universal Garden of Biological Phenomenon. The instant of sequence in space, when water was created, formed, represent important clue in search of biological manifestations, if, no matter how advanced, more or less.

Speed limits in space, it seems, are not as important as following question: how to get there, at any location in Universe, in a matter of an instant. Human body has certain limitations, which set certain standards in terms of boundaries beyond which, travel seem impossible for human being.

The notion of so called grid, woven grid in space, seem to be questionable, unless we are facing virtual reality, written on fine parchment with grid on it, made by virtual codes. Yet, who or what would write virtual reality... .

Universe inherited Intelligent Design of molecular progression, de facto. Subsequently, Human inherited intelligence from Intelligent Universe, senses emerged. Next step of progression is the phenomenon of awareness. What than would be the subsequent next step in terms of logical inheritance. It's possible that the higher manifestation of Intelligent Design, because Intelligent Design progresses in a never ending sequence, according to *The Paramount Law of Transformation.*

Can we visualize above progression ? Yes, as of today, the highest form of progression in Universe is awareness, than the subsequent step is a higher level of awareness, defined profoundly by energy, than a classic molecules.

Existence progresses from simple unaware forms of existence, into more advanced forms of molecular existence, which in turn, are progressing toward transformation from molecular into biological, and subsequently awareness. Next step of progression is projection of energies, due to the fact, that physical, molecular existence has profound limitation, including awareness, expansion of advanced, higher level of logic, as well as an ability to travel through space. What is beyond that, it seem like an abstract, yet, *The Paramount Law of Transformation* is proving that has not limits, by the definition of logical progression ... .

Model of the Universe. Progression of Intelligent Design.

- Space filled with matter, including „cold light" (dark matter), fragmented into simple particles. Space is inanimate.
- Universe is revived through powerful event associated with energy and temperature and interaction between hot light and cold light.
- Expansion of the Universe.
- Formation of more complex physical formations and energies, yet, varieties, are simple in complexity.
- Progression toward Water formation.
- Biological existence emerges, blended from physical particles, elements of Universe, as we know it.
- Senses, a subsequent, profound step in molecular progression.
- Awareness, a revolutionary development, clearly miraculous event, progression of Intelligent Design from simplicity into complexity of projection.
- Transformation of awareness, by expanding a frame of awareness. Multilevel horizon of events associated with higher understanding and perception.
- Transformation of awareness into more advanced formulation of existence, energy, based on density, as well as subsequent logical step, according to The Paramount Law of Transformation.
- Transition of awareness into more complex states of energy.

Human is a biological manifestation of blended energy, electricity, molecules, electro-magnetic field, etc. Transformation of simple molecules and fusion of elements, by forming energies, represent progression of Intelligent Design, from point „0" in space, to the present state of sequence. Expansion of awareness (frame of awareness) is also associated with an ability to interact, through powerful energy of awareness into the physical world. Countless scientific, as well as social interactions between people, are proving progression of awareness, and an ability to penetrate into the physical world, including an ability to subordinate physical world.

After all, Universe, the sequence, was initiated by energy, and this process is progressing toward more advanced mutations of energy. The power of thoughts, the power of human mind, possibilities, is progressing from mode of „receiver of data" into „projection of data". Yet, we don't understand the mechanism, and how to initiate this process, but above phenomenons, are recorded throughout the history of human kind.

Human abilities are limited, as of today, very humble, yet, promising possibilities. Reading of human thoughts is perhaps, at the moment, feasible anticipation, yet, possible to achieve only by few on the planet. Our intuition is telling us, that human has the potential, enormous, and this potential was initiated once Universe was created, and the potential of progression as well. Universe is a projection of possibilities, very logical in nature, according to *The Paramount Law of Transformation.*

Cosmos vs Possibilities of Universal Intelligence

Perhaps every person on this planet, who has a TV set, experienced snowy „noise", which is, de facto, ancient cosmic microwave background radiation (CMB or CMBR). While watching this noise, you see precisely a moment when space, filled with molecular potential, waiting to be alive once again by impregnation through Intelligent Design. As of today, science is indicating, that the microwave background radiation is as old as the Universe, because thermal radiation is the product of the „Big Bang", as few other phenomenons associated with Intelligent Universe, which emerged during formation according to *The Paramount Law of Transformation.*

While watching CMBR, and later projected image (TV show), I realized that two shows, CMBR vs Pink Panther indicate progression of matter, energies into the complexities, which reach beyond necessities of existence, and evolve up to the moment, when sense of humor, comedy becomes a reality. Comedy represent luxury of molecular progression and surely more advanced than we realize.

Does Universe has a sense of humor ? Certainly, if entire spectrum of living matter and energies, progressed to the level of pleasant silliness. Dear uncle Albert, you've asked about God's thoughts once. God's thoughts, are logical, intelligent projection based on Universal Intelligent Design of matter, progressing, de facto, until reaches a state of humorous manifestation, which is compatible with The Law of Possibilities according to *The Paramount Law of Transformation*.

Every scientist would agree, that sense of humor, The Ministry of Silly Walks, for example, is not the most important quality of molecular progression, yet, if progression reached the stage of sense of humor, than most certainly, became essential in biological, sensual, aware molecular existence. Sense of humor, in National Lampoon's Christmas Vacation, is proving that from matrimony between smart molecules, as well as intelligent energies is emerging projection of important and unexpected possibilities, up to the point of hysterical laugh in Naked Gun. It seems that seriousness of molecular existence was not the destiny of progression, but just a stage toward The Mask, more advanced and complex in simplicity transformation, which perfectly illustrate Intelligent Design, most eloquently projected by Louis de Funès. He is, in every aspect of molecular progression, projection of matter, up to the point of transformation, which squeeze tears from one's eyes, while watching the genius of Universal Possibilities.

Fire and Water

Water and fire can not merge. Of course it's not true, when it comes to the progression via Intelligent Design, according to *The Law of Possibilities*. Human is the result of perfectly blended, water and fire.

Progression of Universe

From intelligent, logical patterns emerges molecular machine, apparently biological, including awareness. Intelligent Universe is assembling variety of „machines", yet, with a profound sense of proficiency, elegance, harmony, symmetry, logic and beauty.

Progressing Intelligent Design, through assembling of machines, is navigating toward awareness, and subsequently more complex awareness, awareness which is evolving into the energy, sophisticated progression of molecular projection from a point of initiation of sequence. There is, certainly, nothing more advanced in molecular world, than awareness, yet, visualization of progression in human reality, into the energy of aware existence is profoundly appealing and real, yet, touches abstract, same as projection of CMBR, which we pursue to understand.

Compatibility of Opposites. Intelligent Design.
Equation of Love

Today I've read interesting chapter about mathematics, as well as the equation of love. Apparently I came up with simplified model of the equation, 12 years ago, while living in USA.

As of today, love represent immeasurable quality, yet, we can distinguish two important properties: object 1, object 2, and the result does represent physical quality, de facto, equally compatible with the initiation of the Universe, as we know it (Big Bang), as well as human relations, personal and social, according to *The Paramount Law of Transformation* (Biological System of the Universe).

Object 1, and object 2 represent precise fragmentation of the universe, as well as compatibility of opposites: hot light vs cold light, male vs female.

Equation of Compatibility of Opposites. Intelligent Design :
(Equation of Love)

$$1 + 1 = \infty$$

Object 1: initiation of sequence and subsequent formation of the Universe.
Object 2: space filled with inanimate matter, cold light.

∞: infinity (procreation of matter and sequential progression once sequence is initiated according to **The Paramount Law of Transformation**).

Love represent tangible physical property, which include: projection of a specific kind of electromagnetic energy within and beyond human physical and emotional (emotional = energy) frame (often called aura), electric stimulation of brain waves, and projection of energy, to mention just a few.

By tangible, I understand that if physical phenomenon occurs, than is projected in space and progresses within and beyond (interacting), within specific environment, as well as frame of sequence.

Certainly algorithm of love would reflect extended complexity from the source: $1 + 1 = \infty$, yet, first step has been made.

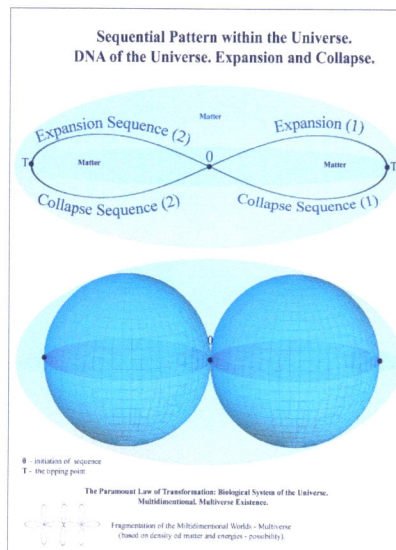

Sequential Pattern within the Universe.
DNA of the Universe. Expansion and Collapse.

Symmetry vs Antisymmetry

Symmetry, the fundamental quality of symmetry at any given frame of sequence, is compatibility. Compatibility of Opposites is fundamental according to ***The Paramount Law of Transformation.***

Important, subsequent question is as follow: what than is the Antisymmetry ?

Antisymmetry is the Expansion of Sequence 1 vs Expansion of Sequence 2, as illustrated in attached example.

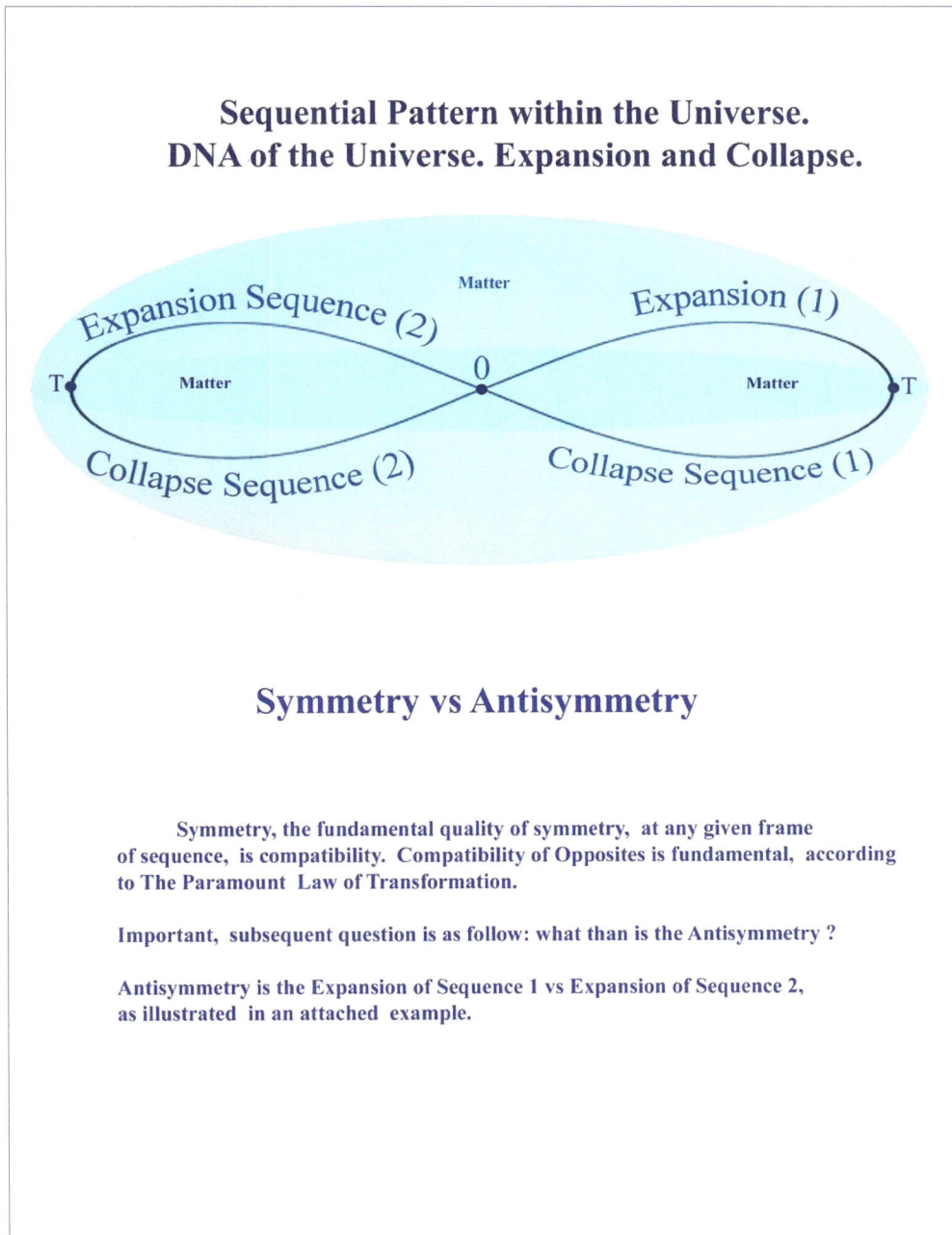

**Sequential Pattern within the Universe.
DNA of the Universe. Expansion and Collapse.**

Expansion Sequence (2)   Matter   Expansion (1)

T   Matter   0   Matter   T

Collapse Sequence (2)   Collapse Sequence (1)

Symmetry vs Antisymmetry

Symmetry, the fundamental quality of symmetry, at any given frame of sequence, is compatibility. Compatibility of Opposites is fundamental, according to The Paramount Law of Transformation.

Important, subsequent question is as follow: what than is the Antisymmetry ?

Antisymmetry is the Expansion of Sequence 1 vs Expansion of Sequence 2, as illustrated in an attached example.

Standard Model vs EQ Wave Super Energy (brain).

Universe in its simplicity in complexity, profoundly manifest gradation of density, from heavy to less dense, until molecular structure is transformed into the energy, where density becomes so minimal or exists under specific circumstances.

Yet, the most advanced of molecular progression, since initiation of sequence, is human thought. Human thoughts project energy, and are recognized by different bandwidths:

- DELTA WAVES (.1 TO 3 HZ)
- THETA WAVES (3 TO 8 HZ)
- ALPHA WAVES (8 TO 15 HZ)
- MU WAVE – (7.5 – 12.5 HZ)
- SMR WAVE – (12.5 – 15.5 HZ)
- BETA WAVES (12 TO 38 HZ)
- GAMMA WAVES (32 TO 100 HZ …)
- **EQ WAVE SUPER ENERGY (EQW)**

Perhaps Gamma Waves represent a threshold not only to understand and grasp awareness, but reaching far beyond perception, which we call reality, as of today. It is also possible, that the most powerful force of energy, is the combination of all brain waves.

Personally, I believe that human brain is able to produce Super Wave Energy (EQW), which surpasses all known properties of brain waves, to date. **EQW Super Wave** is, the facto, an energy, which interacts with physical, more dense elements, as well as existent energies within physical world. In addition **EQW** represent an ability to project its quality, through transformation of physical world.

Can energy, **EQW** be intelligent ? Certainly.

All physical manifestations within the Universe, as we know it, since the initiation, are profoundly intelligent through molecular, and energetic data, embedded within, which transform toward more complex, less dense progression of molecular existence. Molecular existence (unaware) is progressing toward biological, aware molecular/energy, and subsequently progressively less dense, until is transformed and reaches awareness, which progresses in a same manner as Universe does, since initiation of sequence.

Molecular progression, in terms of transformation is simple: molecules, matter, energy, since the initiation of Universal sequence is transforming toward biological existence, biological existence is progressing toward awareness and powerful energy, energy, which not only represent tangible properties (such as bandwidths), yet, is reaching a level of energy of awareness.

Phenomenon of awareness is also progressing into the more complex and advanced formulations of energy, perfected, refined, more potent, according to *The Paramount Law of Transformation.* Awareness is the beginning of a new sequence in terms of **Universal Progression.**

Universe is intelligent, yet, advanced form of progression of elements and energies is **EQ Wave Super Energy** which is , de facto, energy of awareness and **EQW** is expanding.

What is awareness ? Awareness is a perceptual ability to experience projection of sequence within specific reality.

EQW, gradually penetrate physical world in an instant. This is the future of human travel, where energy of **EQWaves** (including awareness) is able to change location in an instant.

EQWave Super Energy, aware energy is „able to see, feel, experience" far beyond the status quo of modern perception and possibilities of interactions … .

EQW Super Waves
Aware Energy

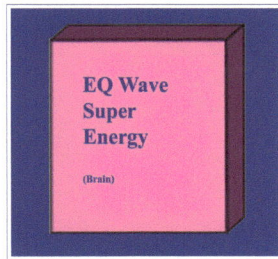

The Paramount Law of Transformation
EQW Super Energy

Intelligent Universe, since initiation of sequence, is progressing into the more advanced forms of Intelligent molecular progression, through transformation of matter and energies.

Intelligent matter and energies represent embedded data of transformation, where one element remains the same, progression into the more complex projections, according to
The Paramount Law of Transformation.

If progression of transformation reaches a state of awareness, than progression of awareness into the more advanced forms of awareness, including aware energy, is logical, natural, and compatible with the entire system of Universal Design, profoundly intelligent, de facto. Human biological system is an illustration of molecular progression into the energy, energy with unique quality, awareness, which is expanding.

System of progression within Intelligent Universe, as we know it, never ends, never reaches boundaries, but is evolving toward refinement and progression from one state into the another, more complex, and de facto, more powerful.

Reality is a state of gradual progression, as well as aware projection within and beyond physical, molecular, energetic fragmentation of awareness. Universe, as we know it, is a progression as well as projection of certain properties. Human brain does function according to the same principles, where Biological System of Intelligent Universe is precise and eloquent definition of Universe, as we now it.

Fragmentation of Universe is compatible with fragmentation of awareness, based on the principle of compatible opposites, for example, where opposites complement compatible opposite into progression and subsequently projection of properties, where projection of progression is manifested by expansion of awareness, same as in Universe, since initiation of sequence.

Expansion of Universe represent identical parallel with expansion of awareness … .

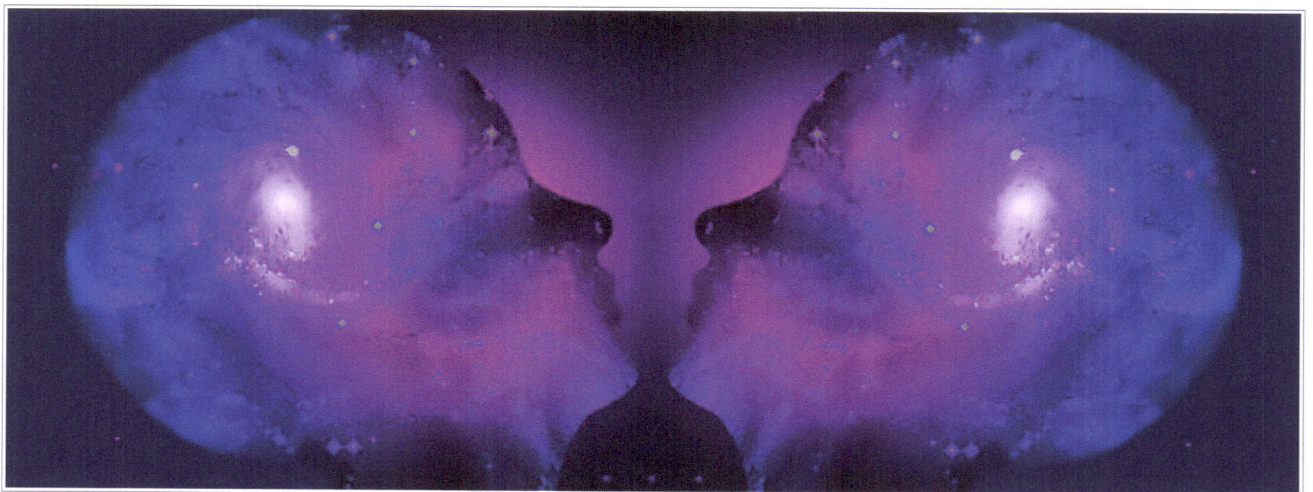

The Paramount Law of Transformation vs Light

Light is navigating the Universe, as we know it, since the initiation of sequence. Light, through all spectrums, is interacting, and shaping the world. Without light, and all known, and unknown spectrums of light, emerging through progression of transformation, Universe would remain as space of inanimate potential, similar to cryogenic state.

Light is the begining of sequence, yet, progression of transformation is beyond human perception, as of today, and perhaps, will never be completely revealed, but progressively expanded.

Light is navigating the world, light transforms matter into the energy, and subsequently into the more refined projections of transformation. Light, as we know it, is the begining of the potential.

Light is a living physical phenomenon, yet, the most unexpected development, is that light is navigating the world toward awareness, and awareness is progressing into the more expanded frames of awareness, less dense, and more powerful penetrating the physical world.

Beginning of the potential, is also observable in biological existence, including aware, from initiation sequence, progression of data available within DNA. DNA represent data, which is dependent on existence of light, as well as emotions. Same quality is manifested in physical, living, yet, unaware physical space.

Parallels between light, and biological progression, including awareness, progression of data, toward complexities, which become aware energies, is not only unexpected, but is proving how profoundly Intelligent Universe is. *Intelligent Universe* is progressing toward progression of intelligence, and each state is the beginning of the next, more complex and powerful. All of that is embedded within Universe and the man. Man is the ultimate, and miraculous blueprint of Intelligent Design, which progressed from Intelligent simple, to the Intelligent perpetual ... .

Light is superbly intelligent, light navigate the world, Universe as we know it, according to *The Paramount Law of Transformation,* beyond available frame of awareness. Light transforms physical world into energies, and progresses into the more advanced aware energies. Each man is different, each galaxy is different within the Universe as well.

Progression of transformation within the Universe, through superb qualities, and spectrums of light, stimulate physical world, and mark the begining of the new projection, as well as potential, where each potential becomes the beginning of the subsequent potential ... .

While thinking about it, persistent singular thought is emerging in my own awareness:
I believe in Intelligent Universe, garden of matter and energies, predesigned, until you, as well as myself, reaches the state of aware manifestation, up to this day.

Believing is a profound physical state, without it, progression of awareness is incomplete, same as a society would be without sense of humor, which becomes essential, from the moment of its manifestation.

Believing in the Intelligent Universe is a logical progression of mind, which had its beginning 13.75 billion years ago (more or less), when sequence was initiated toward progression, which reached the state of something between yesterday and tomorow.

I accepted the obvious, and I'm glad to be aware of my roots, roots of progression and projection of Intelligent Design. I'm glad, I can share with you those thoughts.

Another important development is progression within the same projection (human being for example), yet, always manifested through biological uniqueness. If this quality is present, than we can anticipate variations of aware energy, which become similar, yet, never represent replication, but progression of uniqueness, intelligent automation, if you will. And this quality also indicate intelligent progression within the Intelligent Universe. Intelligent awareness is too intelligent to replicate itself, but progresses into the unique diversifications.

Similarities between Universe, living, unaware and biological, living and aware, is obvious. Planets, galaxies, stars are never the same, yet, unique and diversified in shape, color, manifestations of unique personality, same as apples.

Once you accept the notion that man is a living, aware Universe, embedded within the essence of molecular automated progression, with regard to Intelligent Design (DNA represent automation of progression), as well as energies, than the world, as well as man within, will reveal its secrets eloquently.

Fragmentation of science, between man and universe, as we know it, is non existent, but is a whole, fantastic mosaic of interdependence and interaction, where the essence of the mosaic is compatibility of opposites, not by chance, but intelligent design.

Universe, as we know it, including light, is progressing since the initiation of sequence, and is showing its qualities through progression of energy, and will reach more advanced formulations, projection of energy, including aware energy, which has its source in interaction between hot light and cold light in space, since the intelligent energy of light penetrated space, dark, emotionless, cryogenic.

Light progresses from interaction of compatible opposites, yet, at some point, dark energy will become obsolete.

I believe in Universe, Intelligent Design, living matter, up to the progression, where awareness becomes energy, and energy becomes awareness ... .

The Paramount Law of Transformation vs Art

When we analyze phenomenons of the Universe, important aspect in this process, is to accept physical progression, up to the point of projection of certain events. In previous chapters, I wrote about a sense of humor. Sense of humor is the progression of matter, until it reached a stage, where a „sense of humor" is the „accessory" of Intelligent Design. Yet, if it does manifest its molecular, energetic mutation, event, than it becomes a necessity, in terms of progression of biological existence, subsequently awareness.

Matter, as well as energies of the Universe, are progressing toward refinement through transformation.

Another essential quality, which emerged through projection of intelligent matter, and energies, is awareness progressed toward **creativity**. Can we say accesorized, and not essential ? Creativity, expanding sense of aesthetic perception is parallel with expansion of the Universe, as we know it.

Creativity is profoundly important, because is mirroring essential quality of the universe, yet, more importantly, the dynamics behind initiation of the universe, as we know it, including intelligent design, progression toward biological existence, phenomenon of sensual perception, awareness , which is projecting new frontier of transformation, this time, intelligent, aware energy, energy, less dense, and increasingly potent, in terms of interactions with the physical world.

Creativity also manifests a path toward new sequence, including creation of reality, either virtual or physical, molecular. **Phenomenon of the creative potential** is parallel between the universe, as we know it, and the man … .

Human and the Universe, are parallel if not identical, in terms of primary functions, yet, the difference is complexity.

The Paramount Law of Transformation vs Light and the Truth

In simplicity of complexity of existence, truth represent absolute, in terms perception. Yet, truth is and can be defined with 100 % accuracy.

According to **The Paramount Law of Transformation** the existence of light, as a creative force, as well as universal navigation within the universe, as we know it, is the ultimate truth. This isn't a philosophical statement, but proven scientific fact.

Light, all spectrums of this miraculous phenomenon, is progressing from initiation of sequence to aware energy, and is parallel with expansion of the Universe, as we know it, as well as expansion of frame of awareness, intelligent, aware energy within biological existence.

Parallels between man and the Universe is complete, and complement each other in every physical manifestation, projection of matter, energies, and apparently this process is progressing until it reached intelligent, aware energy.

Progression of aware energy is navigating toward a tip of the triangle of molecular progression, as well as energy, which define physical reality … .

Light, according to **The Paramount Law of Transformation,** represent ultimate truth … .

Light and truth, reflect progression of science, as well projection of possibilities, and possibilities is the perpetual progression, infinity, de facto… .

Light through physical, as well as biological properties, reveals truth in terms of possibilities with embedded data within molecules and energies … .

Light is the truth, and the path leading to variety of truths, physical phenomenons, existent through automation of progression. Without a light, and its spectrums, truth is non existent, and apparently world is non existent.

Light is the beginning of all truths, known and unknown, including intelligent, aware energy within human mind … .

Light, spectrums of light, represent science of truth, reality and the progression of existence, profound phenomenon, which define everything, through projection of infinite possibilities, including aware, intelligent awareness … .

The Paramount Law of Transformation vs Light vs Temperature vs Biological Blueprint

Biological blueprint of Universal existence, Intelligent Design, indicate another important quality with regard to the initiation of the Universe as we know it. Four seasons illustrate a system of molecular transformation as well as progression toward the sequence of physical projection from cryogenic state.

Universe, before initiation sequence was in cryogenic state, with vast possibilities, according to the law of the potential. Initiation sequence within frozen space, filled with all essential elements, was exposed to the energy of light as well as temperature, where light along with temperature progressed toward interaction between hot light and cold light as well as expansion of the universe.

Human biological system is totally dependent on light, as well as temperature. Climatic zones, within biological blueprint is parallel with Universal Intelligent Design. Interplay between hot light and cold light, gradation of temperatures, define predesigned reality toward progression of existence, including projection sequence, awareness, de facto.

Biological blueprint of the universe, is parallel with another anticipated by careful observation quality, engineering of data, which is seed. Seed either illustrate entire Universe, with essential and complete set of data embedded within, or certain element within the universe, as we know it. Seed is data, and the progression of potential is dependent on light and temperature.

Seed represent miraculous potential, yet, the potential is dependent on engineering of circumstances, precise set of data in order to progress and project its varieties within the universe, as we know it.

Biological blueprint of the universe is eloquently projected within human existence, where light and temperature is of paramount importance, along with all essential functions.

The primary engine of the universe, according to **The Paramount Law of Transformation** is space, data, which include transformation sequence, light, temperature, carbon and other essential elements, which progress into other molecular, and blending energies.

Entire system of the Universe, as we know it, is dwell within human body, elements, as well as progression of aware energies. Yet, the engine of existence, and an ability to transform is brilliantly intelligent.

Human is the blueprint of the universe, eloquent, intelligent projection … .

The Paramount Law of Transformation vs Idea

Idea is inherited in existing sequence, as the projection of progression. Idea manifests creative force within structure of data, since the initiation of the Universe, as we know it. Idea is not an invention of human, yet, the idea (data) invented man, through the preprogrammed sequence, molecular progression, expansion of the universe, biological existence. The progression of human sophistication, including idea, is parallel with sophistication of progression of the Universe.

Universe with the entire data embedded into the molecular structure and energies, including human DNA, intelligence, abstract thinking, and magnificent awareness, illustrate sequence and apparently sequence is in idea or progression of fragmentation of destiny, according to **The Paramount Law of Transformation.**

Idea manifests **The Law of Possibilities** or perhaps, The Law of Possibilities is an Idea. Yet, both reflect the intelligent Universe, and intelligence is always associated with The Law of Possibilities, Idea, de facto.

All answers are embedded within human functionality, including an idea, which has its source in molecular and electromagnetic energy. Progression of projection is the destination of existence as well as an idea as an engine of data, according to **The Paramount Law of Transformation.**

The Paramount Law of Transformation is an Idea, and an Idea is The Paramount Law of Transformation, de facto, where The Law of Possibilities penetrate all existence, molecular and energies. Existence, biological in particular, reflect variability in terms of structure, adaptability, variations within its own kind, and this is The Paramount Law of Transformation, projected via an Idea through The Law of Possibilities, where every human projects a different mutation of data, every flower, tree, fauna and flora, clouds as well.

Idea is the creative power, hidden behind every mechanism of transformation, according to The Paramount Law of Transformation, manifested in molecular and energetic diversification of projection with regard to reality.

Perception is fundamentally emebeded into the system of sequential transformation. Without a system of transformation, reality is non existent, perception as well, and this notion is proving that the essential quality of existence, including perceptual, is an idea of transformation, manifested in the Universe, beautifully projected within human existence. An idea is also a path, bridge between the source of Universal existence, Intelligent Design and Human being, which has its source in inherited creativity … .

Triangular Structure of the Universe

Idea

The Paramount Law of Transformation   The Law of Possibilities

The Paramount Law of Transformation vs 3D
Biological System of the Universe

The history of the Universe, as we know it, is the history of light, de facto, its journey through space. The history of light is illustrating sophistication of the phenomenon, as a carrier of transition from inanimate matter, transformation sequence into living, unaware and subsequently aware existence.

History of light is, de facto, progression toward awareness, which via creativity is projecting reversed reality, prior to initiation of sequence, because universe, as we know it, represent data embedded into the progression of transformation.

If energy of light can perform only in 3 Dimensional reality, than we can define reality as 3D. Yet, 1D, 2D represent practical rpgression toward simplification. There is no such thing as 0D, 1D or 2D but 3Dimentional phenomenon. Another example is human body, brain in particular (virtual reality is predominantly 3D).

Light, with all its spectrums, is parallel with the history of truth, within 3Dimensional sphere. Time is flat, without 3Dimensional properties, such as transition of data embedded in space, where sequence of progression, according to **The Paramount Law of Transformation** manifest the essence of existence, where time is irrelevant. Practical applications of time are obvious, yet, to understand the nature and mechanisms of the Universe, the notion of sequence of progression within space, is much more precise. God doesn't need alarm clock to know what time it is.

When we analyze the Universe, than the perception of 3D phenomenon, as well as progression toward higher order, where, de facto, biological existence, human in particular, is the projection of higher order, on the foundation of Intelligent Design.

Attached drawing illustrate not 0D, 2D, but 3D, even if is rendered on flat surface)flat is also non existent in 3D projection). Geometrical form can be rendered only by 3D (human mind, brain), which dwells in 3Dimentional space. Universe, as we know it, progressed from 3Dimensional source, yet, any rendition of 0D, 1D, 2D manifest simplification from 3D, and not the opposite. 3D doesn't produce anything less, than projection of 3D, except simplification toward 0D, 1D, 2D . 3D projection evolve into complexities, based on progression of energies, including aware energies progressively less dense.

Leonardo (3D awareness is projecting 3D reality).

3D reality projection is complete, and represent perfection in terms of progression from one form into another. 3D, by the very nature of complexity in simplicity, represent important parallel with regard to **Intelligent Design**, even prior of initiation of sequence, because 3D can not be rendered (cellular automation) by anything, that is less sophisticated.

Automation of biological, molecular structure, including human, also represent 3D rendition of molecules, as well as energies embedded in 3D space. In Universe as we know it, 0D, 1D, 2D is non existent, because entire structure of Universe manifest 3D projection. Molecules, energies are 3D.

Universe is saying: „you can draw any form you wish, yet, every time you project 3D progression". Why ? Because entire system is based on progression of sophistication, and not vice versa, after all, its is profoundly smart universe... .

3D projection is profoundly parallel with human brain, system of visual perception. To study, in depth, how 3D rendition is possible within human awareness, science will be able to reach a new frontier, a sequence before initiation of the Universe, as we know it, took place.
3D sequence is parallel with an Idea, according to The Paramount Law of Transformation ... .

Human brain doesn't draw points or flat geometrical forms, yet, human brain render 3D projection, same as Universe in its entirety, and the source of 3D projection is light, with all known and unknown, to date, spectrums.
If light is the source of 3D projection (data), than anything less is non essential ... .

Rafael Santi

Attached picture illustrate eloquence of logic in terms of 3D rendition within human mind, as well as Universe. Molecules, and energies within the Universe, as we know it, are 3D and render its own image via cellular automation, which is 3D, yet, the source of 3D is Intelligent Design ... .

The Paramount Law of Transformation vs Science of Reality

Universal system, based on **Intelligent Design**, is based on compatible opposites, projected in every aspect of existence, unaware, as well as aware.

Reality is the result of interaction between compatible opposites, which, de facto, project 3D reality.

Reality manifest important quality, fragmentation system within the universe, and at the same time, compatibility in every scale of existence. Compatibility of the Universe vs Man is complete with regard to awareness vs projection.

Science of Reality

Paramount Universal Reality. Reality which defines all realities.

Universal Reality. Initiation of the Universe, as we know it, and projection of sophistication defined by the Intelligent Design.

Biological Reality. Manifestation of higher order with regard to Universal progression.

Reality of Awareness. Human reality, projection of awareness, as well as perception of higher order, and subsequently aware energy.

Projection of Higher Reality. Awareness is transforming into the energy, aware energy within Universe, as we know it, as well as progression toward **Universal Paramount Reality.** Apparently, this stage is yet to emerge through progression of awareness, yet, the system is compatible and complete.

Reality doesn't progress into so called multiverse (yet we could use this term), but represent transformation with regard to projection of progression, or simply put it, automation toward higher order. Projection of Higher Reality, within existing progression of sequence, illustrate unique ability, in terms of sense of aesthetic projection. Higher reality is projected, yet, perception leads higher reality into the desire, as well as science to replicate aesthetic projection within material world.

Molecular Projection vs Projection of Awareness

Sense of beauty is perceptual, yet, the phenomenon of aesthetic perception is projected through molecular automation, based on Intelligent Design, and transformation into the awareness, aware energy of higher order, which is apparently progressing into the desire to replicate higher order, another phenomenon associated with biological existence, most profoundly manifested by man.

Parallels between molecular automation of Intelligent Design, and the phenomenon of aware energy, within biological existence is complete.

We live in a system of projection of higher order, from molecular, unaware existence, which is projecting qualities of higher order, and aware existence, which is projecting perception of higher order. Compatibility of higher order, between molecular, unaware molecular existence, and aware energy, de facto, illustrate existence of higher order.

Compatibility is mirroring molecular reality and aware reality. Mirror images reflect the phenomenon of projection and the phenomenon of perception.

The Paramount Law of Transformation. Compatibility of Intelligent Design. Science of Reality

Projection	^	Awareness
Projection of molecules and energies.		3D projection and sensual perception (brain).

System of aware projection and sensual perception is evolving into projection of perception through aware energy, as well as progression of aware energy into subsequent higher order.

Progression of awareness. Replication of similarities.

Versus
Projection of Awareness. Projection of Higher Order embedded within Intelligent Design.

Human manifest higher order of molecular progression, yet, similarities evolve into individuality within molecular existence, and subsequently navigate toward aware energy, including Paramount Universal Reality.

The Paramount Law of Transformation vs Motion

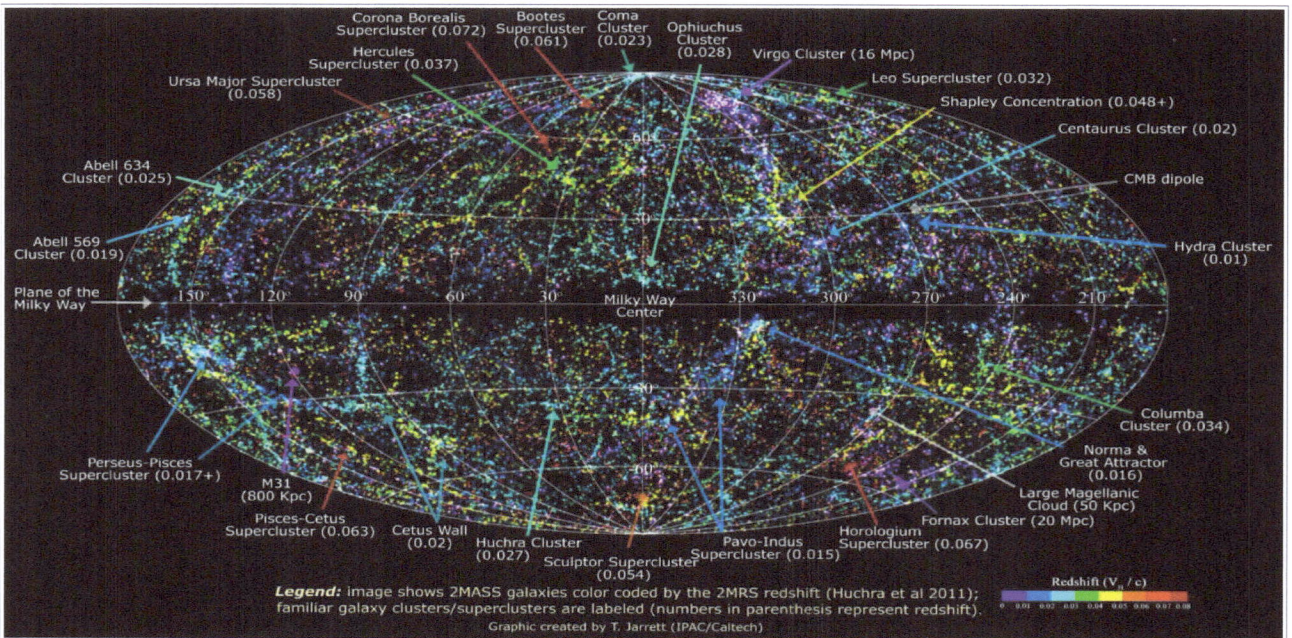

Legend: image shows 2MASS galaxies color coded by the 2MRS redshift (Huchra et al 2011); familiar galaxy clusters/superclusters are labeled (numbers in parenthesis represent redshift). Graphic created by T. Jarrett (IPAC/Caltech)

Modern science is providing unprecedented developments in terms of discoveries. Above map of the universe represent, fantastic scientific achievement.

It's been said that the universe is moving away, expanding at great speeds (for example Milky Way is moving approximately 552 km/s - 600 km/s).

E. Hubble predicted Universal expansion. De facto, Universe is in motion, expansion since initiation sequence took place.

Beside motion of expansion, **Universe is Rotating Around its Axis**. Universe is expanding and rotating until it reaches a state of contraction.

Galaxies replicate Universal system in terms of motion, expansion, rotation and contraction, vortex, de facto.

Galaxies are changing its position, due to the expansion of the universe, yet, more profoundly, due to the rotation around its axis, according to **The Paramount Law of Transformation**. It seems that the Universal engineering is never losing its grip on molecular freelancing. **Universe is scientifically predictable.**

The engine of the Universe, since initiation sequence, is progressing via interaction between hot light vs cold light, and is replicated respectfully in the motion of water. Cold water is circling deep within the Earth in slow motion vs warm water, circling around the Earth, at higher speed, and shallow depths. Both, cold and warm water, are moving at opposite directions.
The engine of the Universe is the interaction between light vs dark matter (hot light vs cold light), including variations with regard to densities of matter and temperatures.

The Paramount Law of Transformation. Light vs Dark Matter.
The Power of Transformation. Parallels. Projection of Intelligent Design.

Universe Pillars of Light: Hubble Telescope.

vs
Earth

vs
Local Universe

All three images, are illustrating **Map of Light vs Darkness** (all spectrums of light) in every scale of existence, as well as Universal Intelligent Design, **Triangular Structure of the Universe:** Idea, The Paramount Law of Transformation, The Law of Possibilities, as well as motion, 3D, molecular projection vs projection of awareness, higher order of progression, Science of Reality, Intelligent Design vs Local Universe … .

The Paramount Law of Transformation vs Initiation Sequence
Creation of the Universe
Divine Toys

Universe, as we know it, is rotating around it axis, interaction between light, and cold light is progressing into the circular motion. Universe, as we know it, is a part of a larger cluster of matter and is rotating.

Initiation of Universal sequence is likely to occur once interaction between hot light vs cold light reaches critical point, a spark, which initiate a new Universal molecular sequence (image of thunderstorm and vortex, above, was taken near Wadsworth, Illinois, USA).

Universe, as we know it, is an integral part of a larger system, as illustrated below, and is always in motion, either internally, externally or both (even if you are standing still, you are in motion). Red dots represent initiation of sequence, once compatibility of opposites reach initiation point.

The Paramount Law of Transformation: Biological System of the Universe.
Multidimentional. Multiverse Existence.

Fragmentation of the Miltidimentional Worlds - Multiverse
(based on density od matter and energies - possibility).

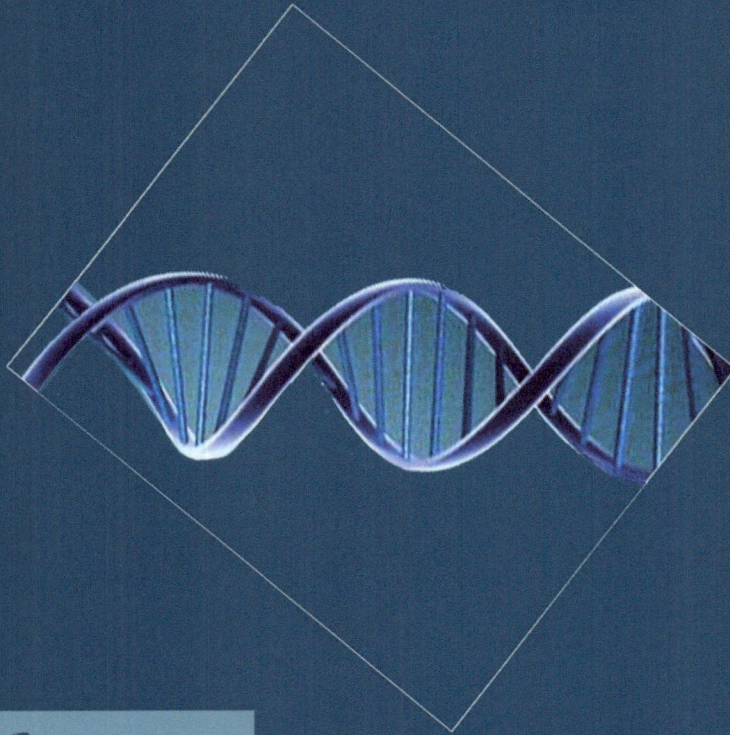

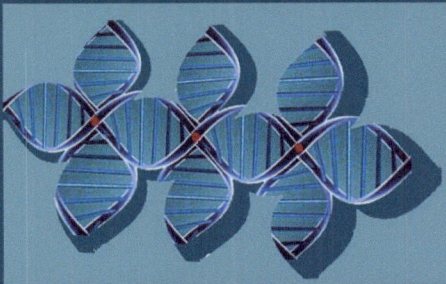

Universe is transforming into the
projection of new universal entities.

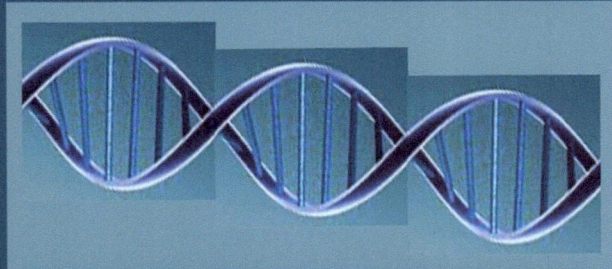

Human DNA

The Paramount Law of Transformation
Biological System of the Universe

Human DNA illustrate, de facto, structural blueprint of the Universe. Human is transforming to the higher level of sophistication, same as the Universe. Universe is an integral part of a larger system.

Human is the Blueprint of the Universe, as we know it.

The Paramount Law of Transformation
Compatibility of Opposites
Definition of Logic

The Paramount Law of Transformation

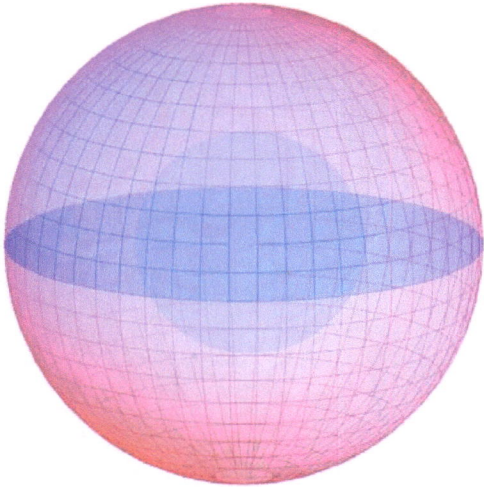

vs

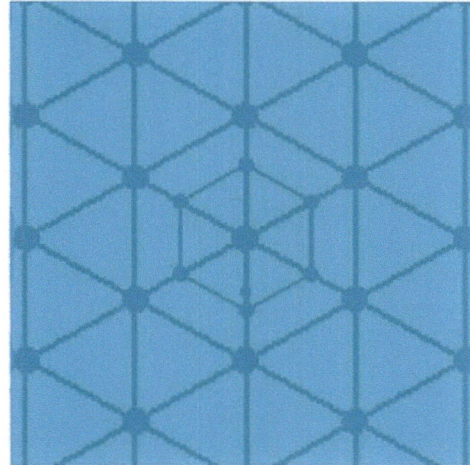

Energies and moleculs.   **Cubicle Science of Logic**

Compatibility of Opposites
Definition of Friction

Universe, as we know it, is rendered in geometrical shapes, de facto, spheres. From subatomic particles to the large clusters of universal bodies, sphere, circle, elipse, vortex, define 3D projection, Intelligent Design.

We all agree that Universe is profoundly smart, subsequently smart universe invented Definition of Friction, as the essential geometrical law, and the essence of the law is sphere.

Universe is projecting quality of frictionless existence, based on eloquent molecular transformation, transformation, which progress into the biological existence, and subsequently awareness, as well as projection of aware energy.

Yet, cube represent compatible opposite with regard to spheere, in this instance, cube project the science of logic, Intelligent Design.

Universe is built on logical cubicle science of logic, embedded into the frictionless molecular and energetic transformation. Fragmentation of the spherical universe is blended by the cubicle logic, Intelligent Design. The entire system is complete. Spherical projection of molecular world is progressing by the grace of the cubicle science of logic.

Perhaps, in the future, 3D projection of awareness will evolve into the 3D spherical logic, aware energy of logic.

The Paramount Law of Transformation

Vortex of Matter vs Transition Sequence. Perpetual Engine.

Vortex of Matter is the engine of transformation of matter and energies, locally, within existing Universe, as well as beyond local Universal composition of Intelligent Design.

Vortex of Matter, the most powerful in Universe, as we know it, is progressing beyond local molecular environment, de facto, is the end the beginning of the molecular status quo, yet, Intelligent Design dwell in a new progression of projection, according to
The Paramount Law of Transformation.

Gravity is common property in the Universe, as we know it, yet, Vortex of Matter can „puncture" through external layer, and initiate a new sequence of progression, „spark" inanimate space, and fertilize new sequence with all spectrums of light and hot temperature.

Biological Blueprint of the Universe is implemented through connected vessels (cardiovascular vessels, cellular biological system, nervous system). By projection of molecular existence, transition of molecules into energies, as well as progression of sequence of physical, and intellectual abilities, Human is the Map of the Universe, within and beyond ... .

The Paramount Law of Transformation

Vortex of Matter vs Transition of Matter

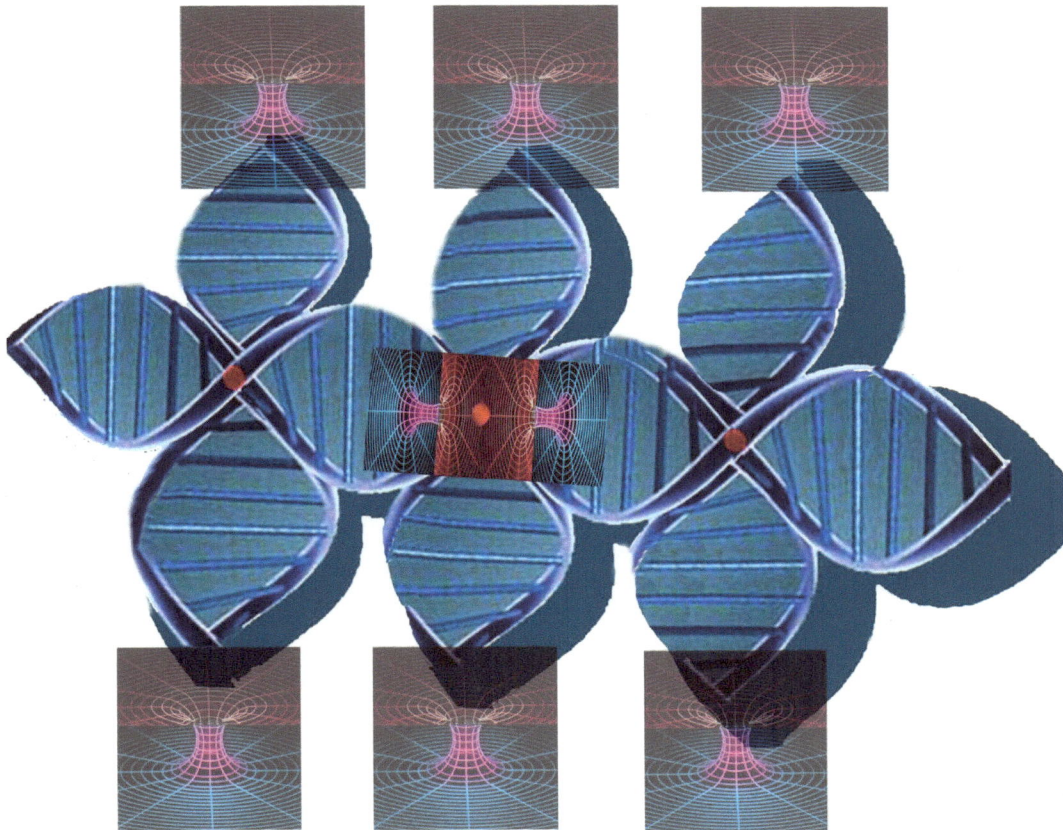

Biological Blueprint of the Universe. Human DNA, as well as Transition System within the Universe.

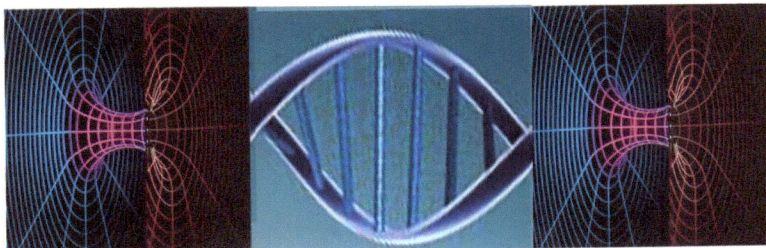

The Paramount Law of Transformation
The History Mind

Mind, besides all available accessorized definitions, profoundly manifest Intelligent Design, where progression from molecular into biological existence, as well as sensual, and subsequently aware energy is projecting a sequence toward higher sophistication, since initiation took place, and even prior to this moment.

Mind, represent empirical reflection of reality, projected by Intelligent Design, where awareness of molecular, and energetic progression is inherited via perception and memory, among other qualities of the mind. Mind, as well as the entire Universal system, represent compatible fragmentation of parallels.

Mind is a system of sensory replication, mirroring, if you will, of the physical world, and at the same time, progression of projection with regard to the higher order of sophstication. Biological blueprint of existence is parallel with Universal Intelligent Design.

All molecules, are „equipped" with memory, designed to perform, according to the certain scheme, yet, with profoundly embedded function of replication, and at the same time, progression of projection (mutations), which is the essential quality of inherited Intelligent Design toward sophistication.

Memory, the essence of retaining and transforming information, data, illustrate paramount role of sequence vs time. Memory represent data reflected within sequential system of progression. Memory is profoundly 3D, by connecting different data within different partitions of sequence.

Imagination, by evoking projections of mutations, with regard to virtual reality, within physical existence, mind, de facto, reflect compatibility with **The Law of Possibilities**, imagination is the virtual exercise of possibilities, limitless by the very nature of imagination.

Human mind, with excessively manifested number of electrically potent neurons is calculated to about 100 billion (more or less). Brain functionality differ from person to person.
This is certainly an example of superb **networking design,** constantly progressing, de facto. The system of networking is parallel with the Universal Intelligent Design. Human, with all its functions, is the Blueprint of the Universe, the blueprint of progression toward sophistication.

Thought, empirical function of the mind, within the system of sequential progression, represent tangible manifestation of the mind, by projection of diversified projections, embedded within intellectual and physical functionality.

Mind, thought represent an energy, electromagnetic energy, de facto, projected through progression from molecular into energies. Evolution of this miraculous phenomenon is compatible with **The Paramount Law of Transformation**. Intelligent Design represent another important quality, it's progressing steadily toward higher order, since the beginning of initiation of sequence, and it seems, Intelligent Design never makes mistakes, never says, „wrong path". All forms of existence, are becoming more complex, sophisticated, including emerging aware energy, progression of the mind toward higher order, which has its source in higher order within Intelligent Design.

The Paramount Law of Transformation
Biological Blueprint of the Universe
Human Eye vs Transistor (input vs output)

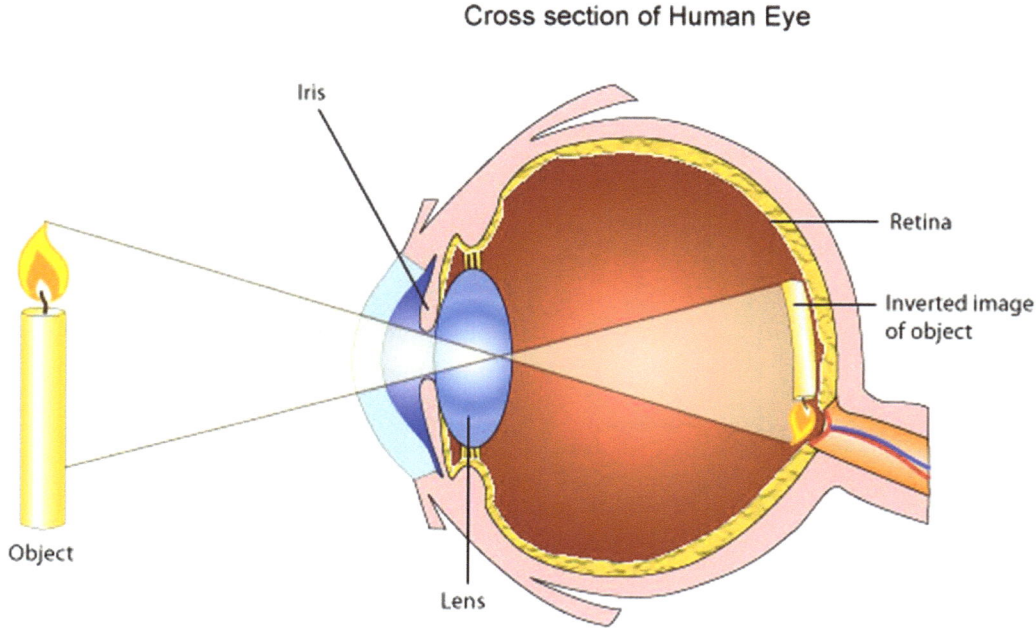

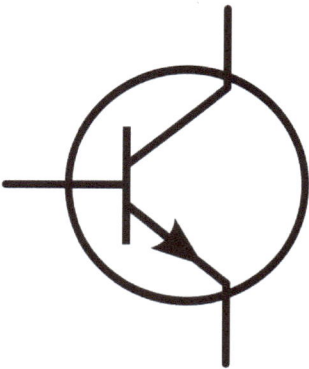

Cross section of Human Eye

Iris

Retina

Inverted image
of object

Object

Lens

Brain. Amplification of data from light.

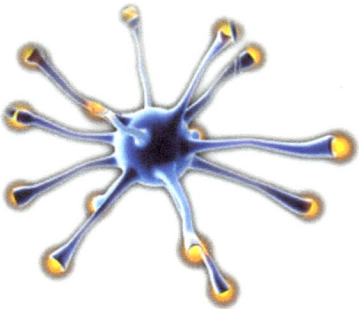

**Transistor
Input (low) vs Output (high)**

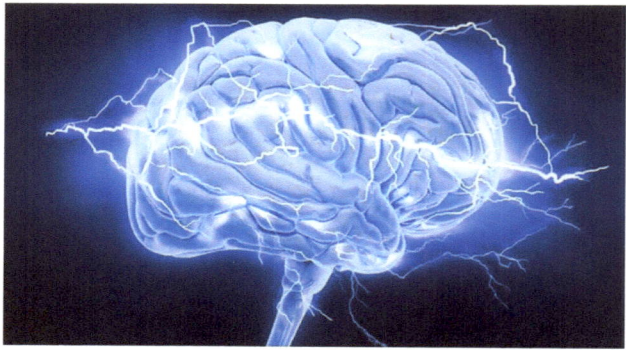

Human Brain: neuron.

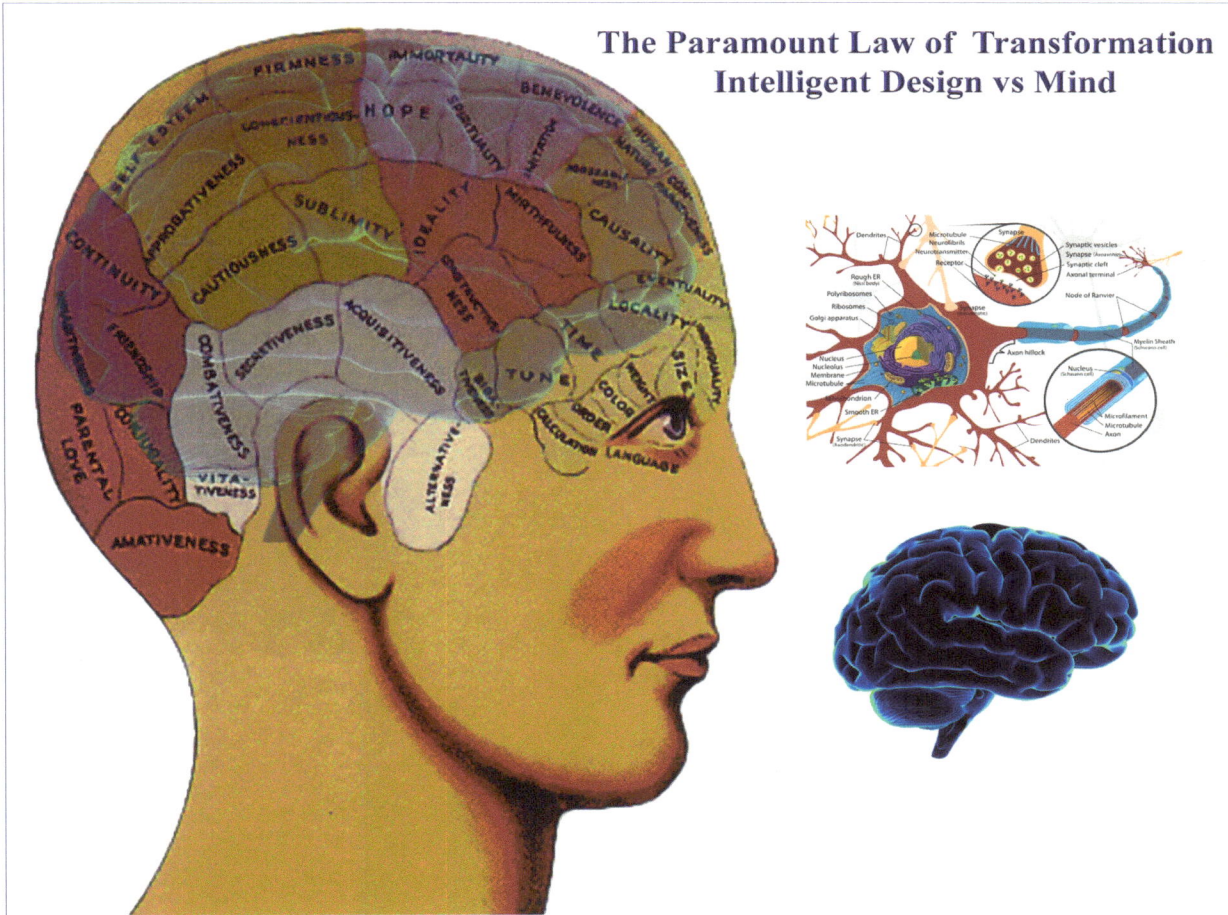

The Paramount Law of Transformation
Intelligent Design vs Mind

Human Blueprint of the Universe
Intelligent Design vs Compatible Fragmentation

Human brain represent interconnected, compatible fragmentation system, nearly exact as the building blocks of the Universe, as we know it.

In addition, in Biological Blueprint of the Universe, amplification of energy is common, similar to transistors, and dwell within the system of transformation, including molecular fusion (low input vs high output). Another example illustrate transition of low density of matter versus high density of matter, and vice versa.

Intelligent Design, and amplification of energy is present in all spectrums of existence, molecular, physical, energies, biological, as well as emerging biological aware energy.

The History of the Mind is as old as the Universe, which eloquently indicate Intelligent Design, based on embedded data of transformation, superbly efficient system of recycling of matter, progression toward higher order, transition from molecular into energy, including, aware. Human is navigating to the source of its existence through transformation. The history of the mind progressed via molecular transition, including biological existence to emerge, miraculously as an aware, perceptual being ... .

The Paramount Law of Transformation

Mind vs Projection of Energy

The Primary function of mind is to process and project data, as well as progression toward higher order.
Mind is progressing into the projection of aware energy. Molecular networking is parallel with emotional networking

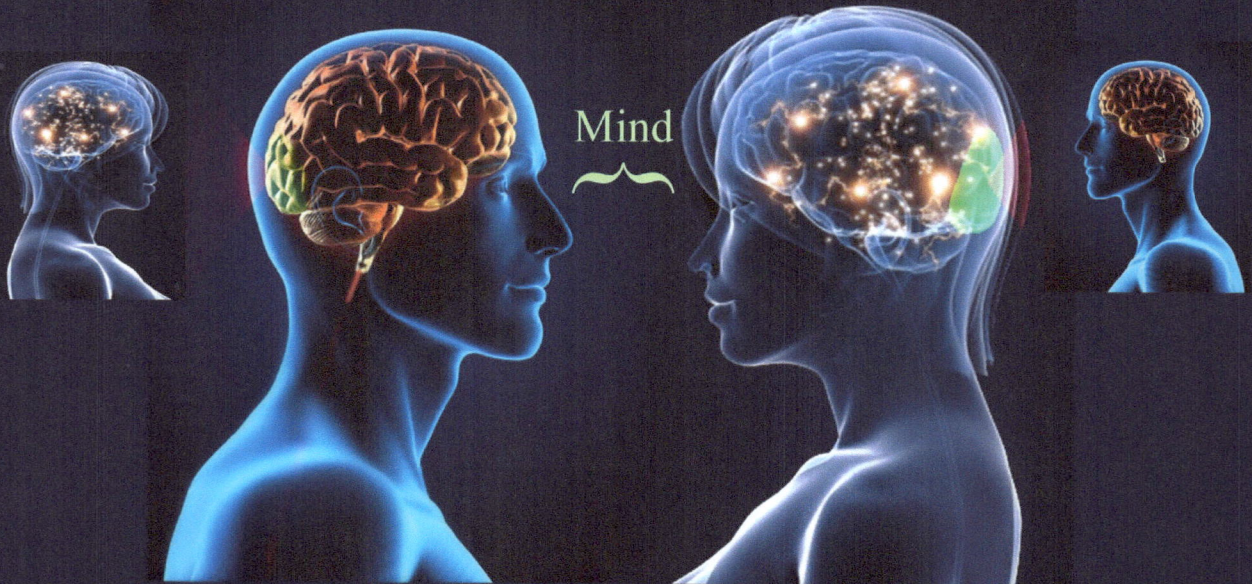

Mind

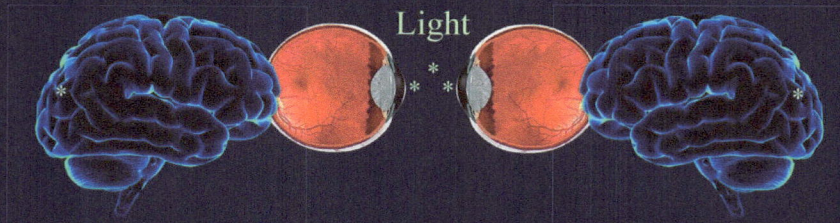

Light

Projection   Projection

Human, the blueprint of the Universe, as we know it. Compatibility of opposites.
Sequence: progression of projection = projection of progression.

Light is, de facto, the primary source of 3D projection, reality, as well as transformation.
Important issue: how far mind is projecting existing reality of reflected light, and how far mind
is eganged in projecting of reality (interpretation), due to the nonuniform perception.
Molecular transformation is progressing into the energy, as the projection of higher order.
Energy is transnformed into aware energy, and aware energy is the subject of progression as well.

Universe
Intelligent Design

Progression of Intelligent Design
Projection of Aware Energy

The Paramount Law of Transformation

Mathematics vs Geometry. Supersymmetry vs Energy.

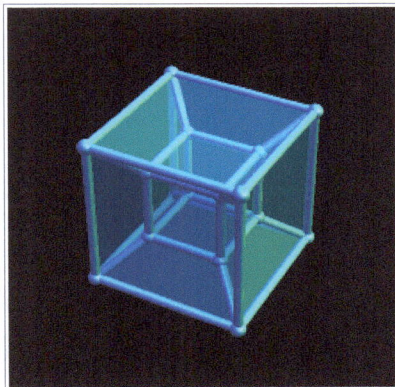

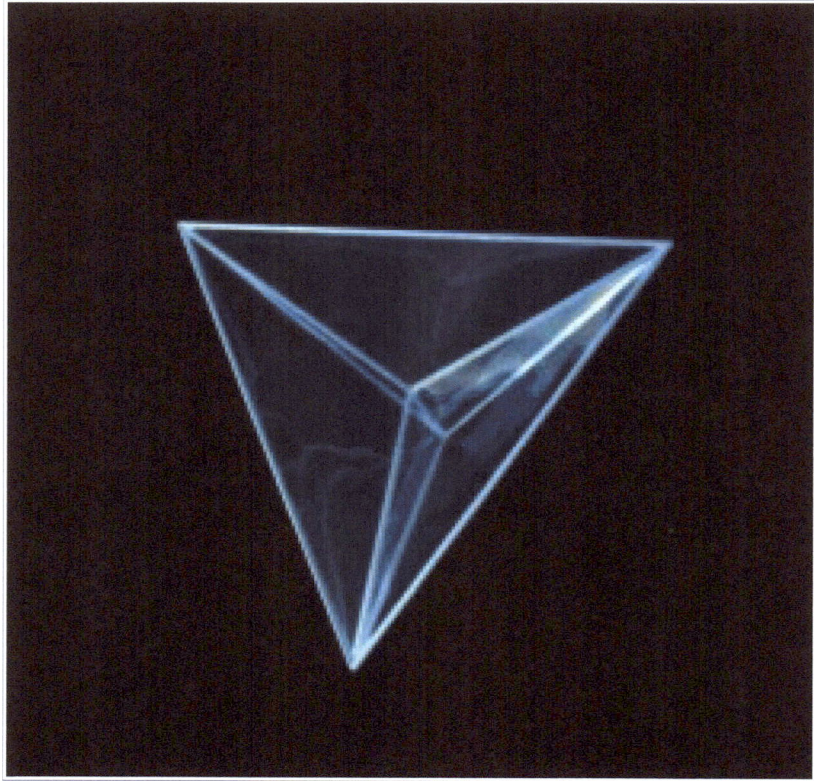

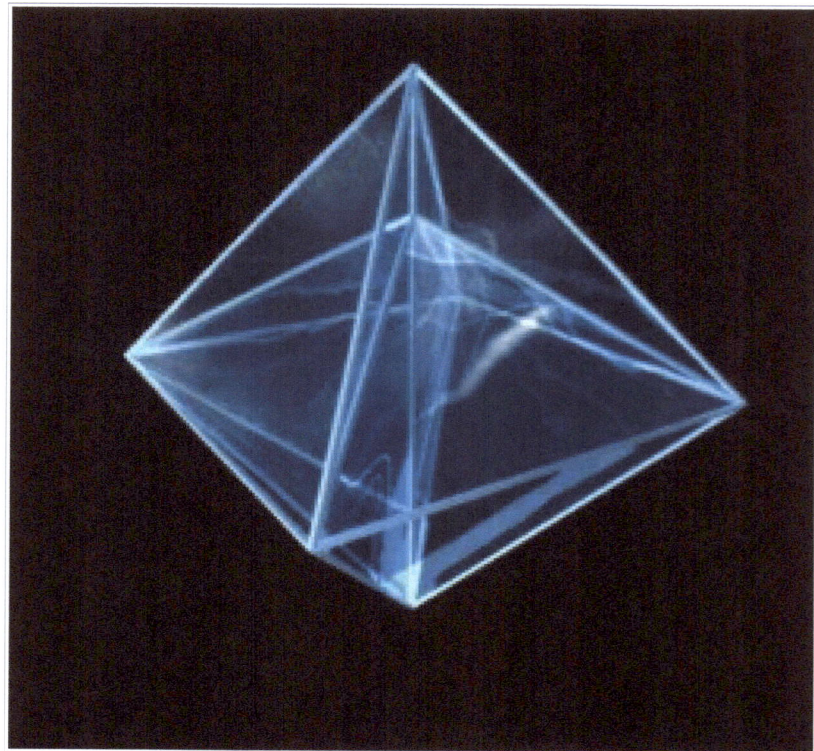

The Paramount Law of Transformation
Biological Blueprint of the Universe
Logical Transition vs Geometrical Universe

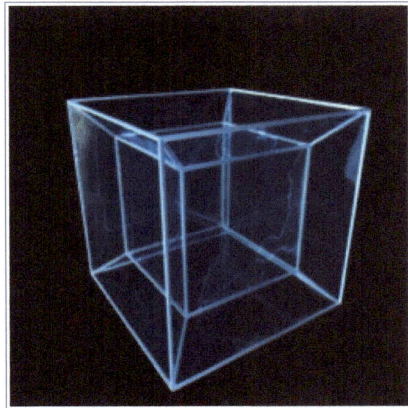

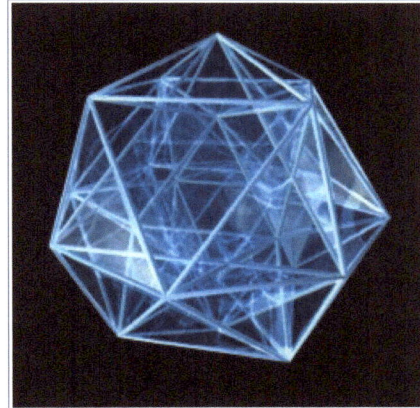

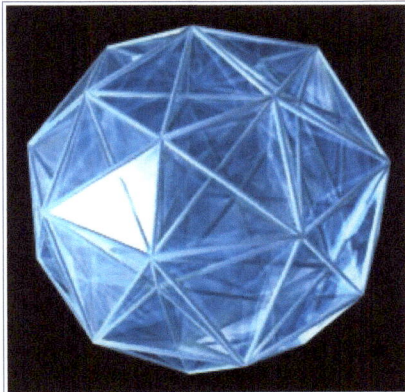

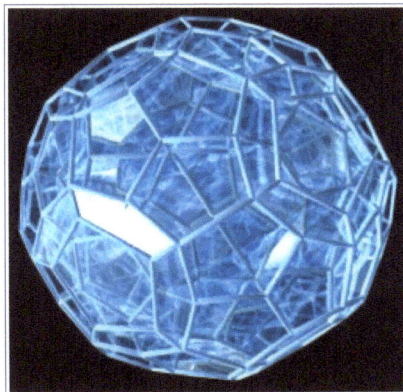

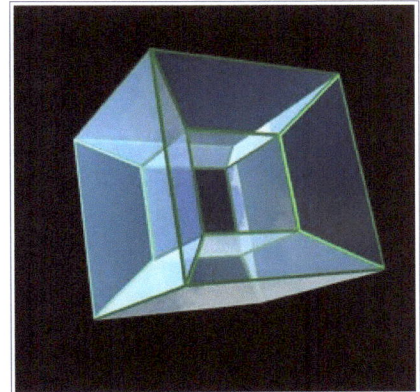

   3D Geometrical Rendition of so called solids, illustrate transition process, progression of projection, which existed since the initiation of sequence, and surely prior to initiation, modern technology, allowed to render transitions.

   Another important aspect of **Universal Transformation**, is that Intelligent Design is performing according to the stipulation paradigm, de facto, simplification: binary system 0 1.

   Simplicity of complexity is, de facto, the most advanced projection of sophistication, not only logically eloquent, but proficient in terms of transformation into the higher order of sophistication.

   Supersymmetry is certainly associated with 3D projection, as well as seamless transition from one projection into another. Take for example geometrical solids, from brilliantly formulated by Plato, Aristoteles, to present time. Logical progression demand that all neccesities of rendition ought to absorb one another or progress in seamless geometrical transitions (in any direction). If they don't, than the sequence of supersymmetry is broken or insufficiently formulated.

Mathematics vs Geometry
Supersymmetry vs Energy

The primary language of progression is, de facto, 3 Dimensional projection, where sequence is engaged in space, through **Intelligent Design**, data (DNA) of transformation into the higher orders. Mathematics, by its very nature is flat, yet, algebraic sequence can be 3D as well.

Mathematics play essential role in scientific process, in quest for knowledge, yet, mathematics can prove fundamental timeless truths, and at the same time, provide platform for theories, which became incorrect. It is a typical scientific process, yet, the objective is to maximize rendition of truths.

Mathematical eloquence can be, and should be rendered into the 3D projection. 3Dimensional projection along with algebraic equations, manifest practical and complete scientific, perceptual fragmentation of data.

Entire spectrum of perception, along with all accessories of mind, including thoughts, memory, logical thinking, abstract functionality, emotions, is rendered in 3D. 3D rendition of perception shows simplicity in complexity and is fundamental in terms of formulating scientific data.

Supersymmetry. Compatible opposites can refer to supersymmetry of progression. Supersymmetry manifest progression into more sophisticated projections of matter, and subsequently energy. According to **The Paramount Law of Transformation**, all components within Universe, as we know it, represent molecular potential, with regard to energy, more or less potent, yet, able to be transformed.

Supersymmetry, within Intelligent Design, manifest essential component of progression, transition, from molecular into energy. In Biological Blueprint of the Universe, as we know it, Supersymmetry is observable in all stages of human development, yet, the potential of Supersymmetrical transformation is still progressing.

Supersymmetry is present in Multidimensional energies and molecular progressions.

Energy. Energy, in nearly all molecular eloquence of transitions, is associated with electricity, yet, more precisely with electromagnetic radiation, either on molecular level, transformed into energy, as well as biological. Energy is the most common, and potent phenomenon in the Universe, as we know it, for example, electromagnetic radiation of the Sun (radio waves, microwaves, infrared, visible light, ultraviolet, X-rays, Gamma rays, thermal radiation and electromagnetic radiation as a form of heat).

The Paramount Law of Transformation
Supersymmetry vs Breaking Supersymmetry

Supersymmetry

Since we explored the Supersymmetry, now is the time to „break" the Supersymmetry. Both phenomenons are algebraically and geometrically interwoven, yet, alphabetically compatible.

3D projection render fantastically potent transformations, and ultimately project the Law of Possibilities with regard to molecular and energetic eloquence of transitions. As I have mention previously, 3D rendition of solids, in terms of Universal Intelligent Design, must be able to transform, and be transformed from one 3D formulation of projection into another.

3D projection is parallel with molecular transformations, as well as energies, due to the fact, that the rendition of geometrical forms emerged from Universal progression of sequence, logical, sophisticated mind, de facto.

Supersymmetry vs Breaking Supersymmetry

Graphic Art by Mr. M.C. Escher "Metamorphosis"
(non commercial use)

Attached illustration project progression of Symmetry, Supersymmetry and prgressivelly „breaking" the Supersymmetry. 3D projection illustrate transformation of forms, with regard to molecules and energy as well. The system is complete: 3D projection, Symmetry, Supersymmetry, „breaking" Supersymmetry via transition into progression of Symmetry, Intelligent Design according to **The Paramount Law of Transformation.** Besides completness, the system of transformation reflect another quality with regard to Intelligent Design, the **Permanence of Aesthetic Projection** within … .

Supersymmetry vs Biological Blueprint of the Universe

**Biological Blueprint of the Universe
Supersymmetry ("breaking" Supersymmetry)**

While studying molecular automation, cellular computation in USA, I realized that the Symmetry progressed into the Supersymmetry, and subsequently „breaking" Supersymmetry. Terminology „breaking" Supersymmetry is slightly misleading, due to the fact, that Supersymmetry is progressing into the new projection, which has its sequential source in Supersymmetry.

Submolecular level of progression is no different, unless, human technology will produce a new, unknown particle(s) or fragmentation of Universal particles. In this instance, human race can reach the situation of disrupting molecular integrity by parallel forces, as Vortex of Matter (Black Hole). Once this happens, the Solar System, Milky Way Galaxy, biological forms of life, would stretch like a glue or jelly, and than, the entire structure of Intelligent Design, would be invited for the ride, (possibly reset).

Supersymmetry is profoundly projected in **Biological Blueprint of the Universe**. For example, brain is symmetrical, yet, fragmented by the design in terms of functionality (right vs left hemisphere), compatibility of opposites with regard to sexes, progression of sequence, since initiation, biological transformations.

The Paramount Law of Transformation
Intelligent Design vs Aesthetic Projection
Biological Blueprint of the Universe
DNA of Molecular Progression vs DNA of Aesthetic Design

Universal Design vs Aesthetic Projection
Biological Blueprint of the Universe

Perception is scientific, no matter if projection of science is rendered empirically or by the intuitive awareness. Countless discoveries claim that intuition was a driving force toward exploration of science, as well as expansion of frame of awareness, including Uncle Albert.

Someone perhaps would ask, why I chose Botticelli's Venus painting as an illustration for **The Paramount Law of Transformation.** It is true that dozens if not hundreds masterpieces would eloquently illustrate Biological Blueprint of the Universe, yet, Botticelli's Venus contain all elements, which are essential, in terms of transformation, Universal progression toward higher order.

3D projection, light, sky, Earth, water, wind (expansion), women vs man (possible relation between hot light and cold light within the universe), waves (energy), five persons, yet, lady on the right side is carrying a little baby in her womb, miraculous fertility of Universe (man vs Venus).

Besides all essential elements within, the entire composition is blending fragmentation into the seamless and brilliantly eloquent phenomenon of beauty, where Intelligent Design compose, beyond necessity of molecular, energetic progression, aesthetic manifestation of the design. Aesthetic perception clearly progressed since the initiation of sequence and subsequently, biological existence, human in particular is the masterpiece in terms of sequence of beauty.

Every time we look into the molecular, structure of the world, every time we look at the amazing images, taken by Hubble Telescope, every time we explore images submitted by scientists, in every discipline, we recognize the higher order of the Universal progression, de facto, the essence of **The Paramount Law of Transformation**, aesthetic projection. By the very act of recognition of beauty, as well as reluctance to accept the opposite to aestheticism, is embedded the essence of the Intelligent Design, as a uniform paradigm of progression.

Molecules function according to the simplification of complexities, yet, the design of progression represent an integral data of the world, including molecular, energies, motion, projection of transformation.

Aesthetic paradigm in not an accessory of progression, but the very essence of existence, without which, world would be impossible to render its possibilities.

Why human is aware of beauty ? Because the higher order or progression exists within each molecule, flow of energy, clusters of galaxies, motion, including vortex of matter, light, since the initiation of sequence. Awareness of beauty reflect our own design, design of higher purpose according to The Paramount Law of Transformation. Yet, the projection of transformation project the essence of beauty even on emotional level.

Aesthetic Paradigm Universal,yet, procreation of beauty, as well as its potency to surprise awareness by subsequent unveiling expanded frame of transformation is by any account, breathtaking and this is the true face of Intelligent Design.

How to define the beauty of Universe ? Perhaps the best illustration is man within, yet, the aesthetic beauty is the most common treasure handed over by Intelligent Design, where potency of functionality is parallel with potency of perception ... .

DNA of Molecular Progression vs DNA of Aesthetic Design

DNA

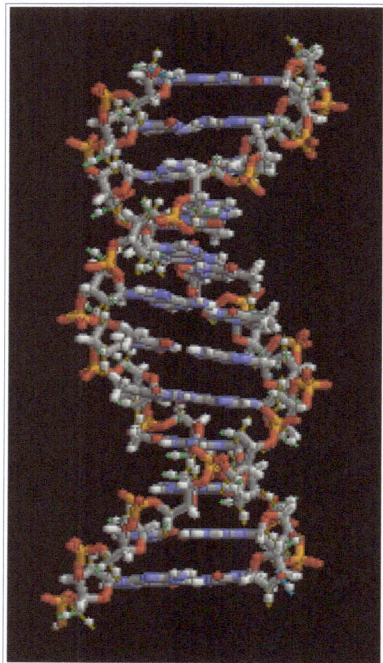

(GIF Wikipedia. Non Commercial Use)

Rodin The Kiss

The Paramount Law of Transformation

Cosmological Constant

Biological Blueprint of the Universe
Biological vs Universal Metabolism

The Paramount Law of Transformation
Cosmological Constant
Biological Bluperint of the Universe vs Biological and Universal Metabolism

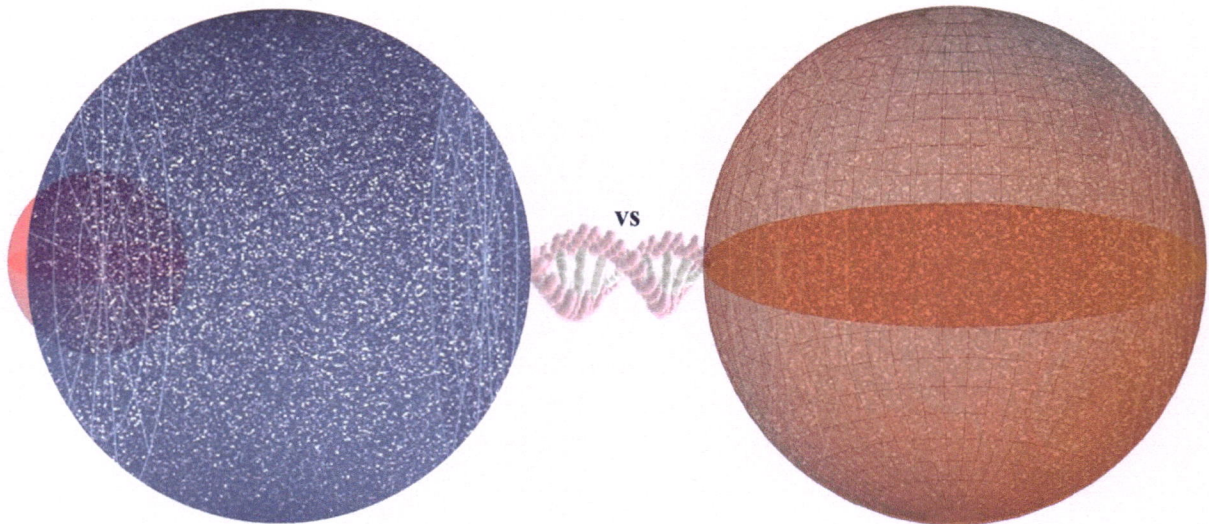

Cosmological Constant
Prior to Initiation of Universal Sequence

Initiation of Universal Sequence (Spin)
Hot Light vs Cold Light (Dark Matter)

Cosmological Constant. Uncle Albert indicated (1917) that Cosmological Constant (Λ) is the energy, density of the vacuum of space, as a static universe. Albert Einstein was right, Cosmological Constant does exist, yet, misplaced in Universal sequential progression. Cosmological Constant is a state of inanimate matter within a space, prior to Initiation of the Universal Sequence (so called Big Bang).

Space is filled with matter, according to the Intelligetn Design, progression of Compatible Opposites, in this instance, Hot Light vs Cold Light. Cosmological Constant represent, de facto, "flat" Universal sphere, without motion (spin), and an ability to transform, yet, predesigned to interact under specific circumstances (Law of Possibilities).

According to **The Paramount Law of Transformation**, inanimate space, is transformed by light, as well as temperature. **Biological Blueprint** of the Universe reflect parallel projection. Human body illustrate equivalent of the Cosmological Constant, transformed by energy of light, electromagnetic field, where temperature is the byproduct of molecular and energetic Universal Initiation, de facto. If human body will be submerged in reasonably hot water, and gradually cool off, human body will produce heat still, long time after event took place, due to the accelerated process, stimuli, heat in this instance. Electroshock (Defibrillation) represent another example with regard to the Biological Blueprint of the Universe.

Biological Blueprint of the Universe
Biological vs Universal Metabolism

Universal Biological Blueprint of the Universe, represent complete parallel scheme with regard **to the Universal as well as Biological Metabolism.**

Metabolism (from Greek μεταβολή metabolē translate as "change"), project transformations within the living organism in order to maintain existence, life, de facto.

Universe, according to The Paramount Law of Transformation, is transforming (changing), equally transparently and eloquently as human being.

Catabolism is the breakdown of molecules to obtain energy: breaks down organic matter and produces energy by way of cellular respiration. **Anabolism,** the synthesis of all molecular composite needed by the cells, which uses energy to construct components. It's all true via projected pathways in Universe, as we know it.

Cellular Respiration (combustion reaction) : reactions involved in respiration, are catabolic reactions, which break large molecules into smaller, releasing energy, as weak "high-energy" bonds, are replaced by stronger bonds in the products.

Respiration is one of the key process where a cell gains useful energy to fuel cellular activity. Cellular respiration is considered an exothermic redox reaction, which releases heat.

Universal existence manifest brilliant simplicity in complexity, as well as compatibility of transformations, de facto, adaptation of universal rules, since the sequence of Universal Initiation occurred. Subsequently Universal Intelligent Design existed before Initiation of Universe, as we know it, occurred.

Biological Blueprint of the Universe vs Parallels

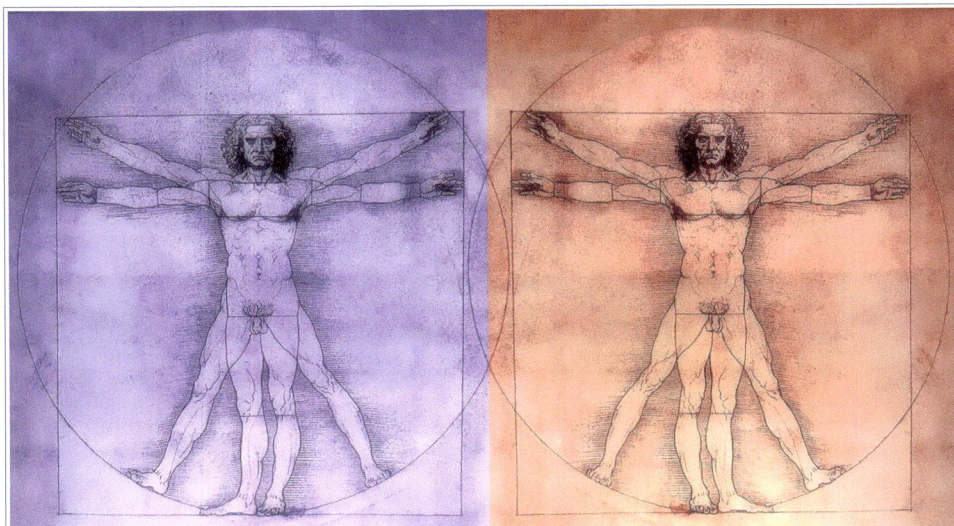

The Paramount Law of Transformation

Composition of Universe vs Composition of Awareness vs Chaos
Higher State of Energy vs Higher State of Awareness

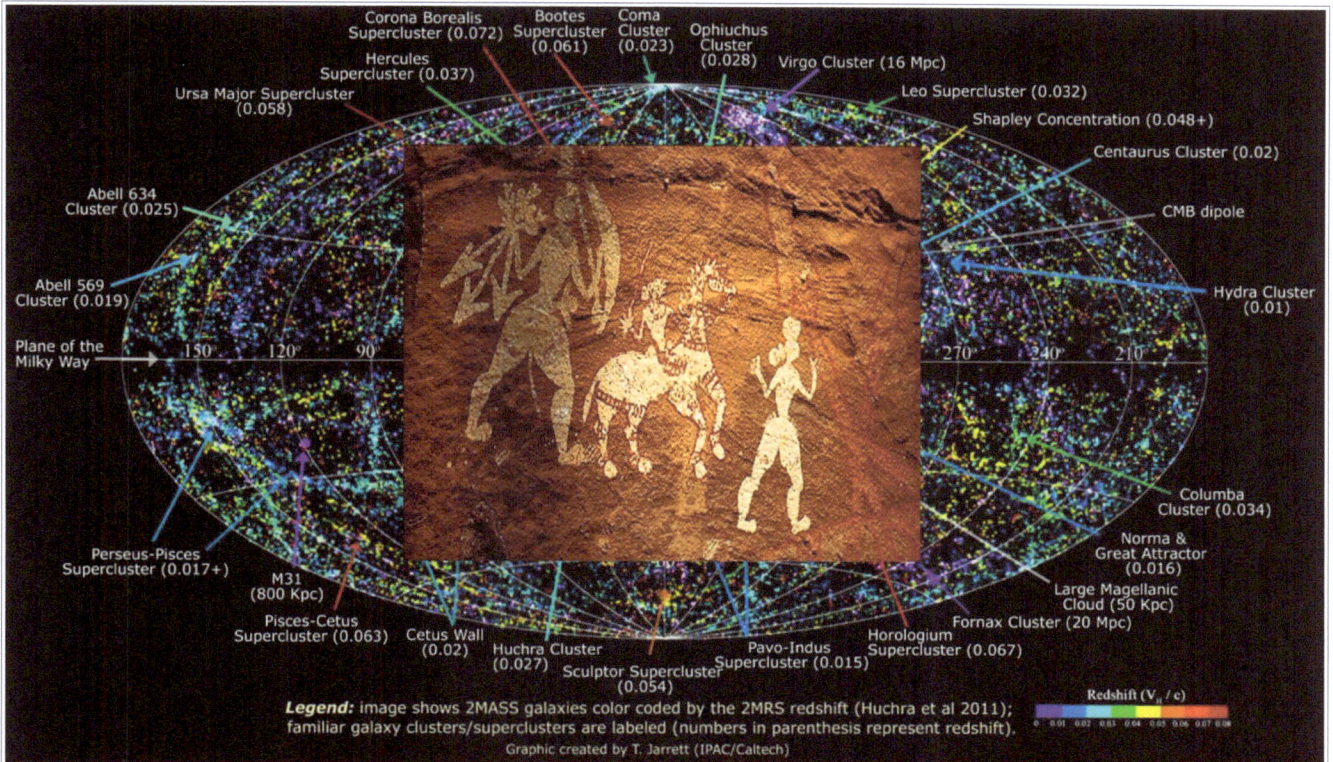

Corona Borealis
Supercluster (0.072)   Bootes
                        Supercluster   Coma
                        (0.061)   Cluster   Ophiuchus
                                       (0.023)   Cluster
                                                 (0.028)   Virgo Cluster (16 Mpc)
Hercules
Supercluster (0.037)   Leo Supercluster (0.032)
Ursa Major Supercluster
(0.058)   Shapley Concentration (0.048+)

                                                                      Centaurus Cluster (0.02)
Abell 634
Cluster (0.025)   CMB dipole

Abell 569   Hydra Cluster
Cluster (0.019)   (0.01)

Plane of the
Milky Way   150   120°   90°   270°   240°   210°

                                                                      Columba
                                                                      Cluster (0.034)

                                                                   Norma &
                                                                   Great Attractor
Perseus-Pisces   (0.016)
Supercluster (0.017+)
                  M31   Large Magellanic
                  (800 Kpc)   Cloud (50 Kpc)
     Pisces-Cetus   Horologium   Fornax Cluster (20 Mpc)
     Supercluster (0.063)   Cetus Wall   Supercluster (0.067)
                            (0.02)   Huchra Cluster   Pavo-Indus
                                     (0.027)   Supercluster (0.015)
                                 Sculptor Supercluster
                                 (0.054)

Legend: image shows 2MASS galaxies color coded by the 2MRS redshift (Huchra et al 2011);
familiar galaxy clusters/superclusters are labeled (numbers in parenthesis represent redshift).

Redshift (V$_H$ / c)
0 0.01 0.02 0.03 0.04 0.05 0.06 0.07 0.08

Graphic created by T. Jarrett (IPAC/Caltech)

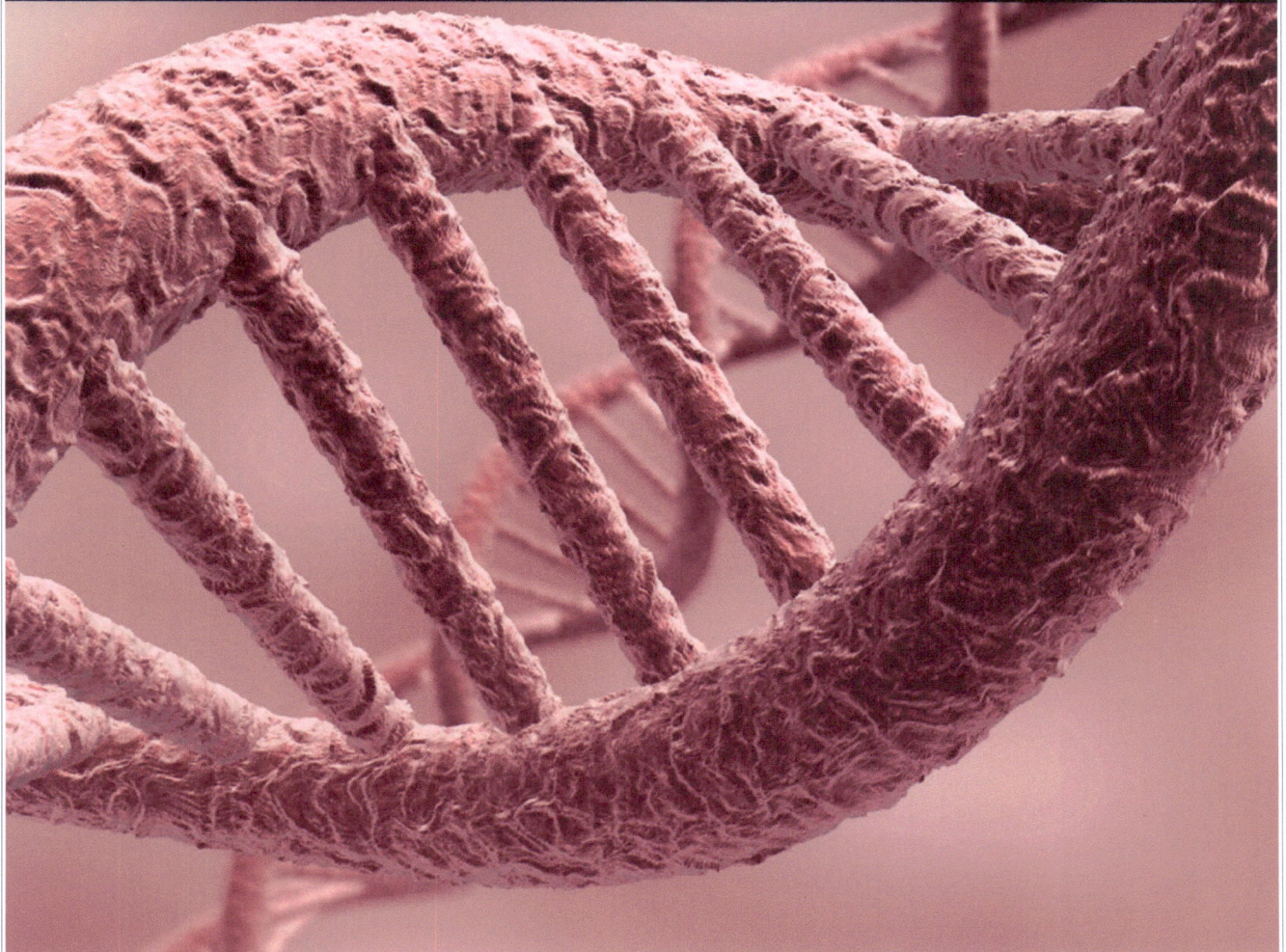

Enormously difficult topics, oscillating between science, as well as scientific intuition, yet, there is enormous data, which illustrate, and provide countless essential logical references.

The notion of chaos is also interesting, because, de facto, is non existent, due to the fact that molecules, all molecular world is profoundly logical and organized.

Chaos refer only to the state where Intelligent Design would be non obligatory, as well as data of progression, yet, we know already that the entire world is showing progression of patterns. Chaos is the by product of sophistication, where Intelligent Design would not properly projected.

The ultimate destiny of Universe is Intelligent Design. Entire Universal environment is logical, simplistic in complexity, and is progressing endlessly.

Molecular world, energies, manifest common quality, since the Initiation of sequence. Universe, with all predesigned data within, according to the Intelligent Design, progress to the higher order. Subsequently biological existence emerged, as well as awareness.

Universe is composed by sophisticated data, information code, which is progressing to the higher order. This is the essence with regard to composition of Universe, as we know it, according to **The Paramount Law of Transformation.** Galaxies represent universal composition, planets as well, even the tiniest molecules are showing a patterns of design.

Human is the most beautiful, and the most sophisticated progression, in terms of progression of composition including, awareness.

The power of universal composition is very much transparent, in every molecular and energetic detail, as well as fantastically blended clusters of matter and energy.

Universe is Composed by the Power of Intelligent Design, including molecular automation, progression toward higher order, sophistication, de facto, interaction between compatible opposites, biological existence, variety of projections, senses, awareness, mind, and finally, aware energy, which is progressing, still, toward higher order of sophistication.

Universe is composed and navigating by the power of Intelligent Design, logical, and at the same time, aesthetic.

How often we hear our own whisper; how beautiful is the world.
Yes, indeed, how beautiful and aesthetically potent is Intelligent Design … .

Design is the projection of logical automation of patterns, within the progression, to achieve higher order, within the perimeter of awareness, and subsequently, creation of a new expressions, yet, compatible with Universal Design.

Human is the projection, de facto, of a magnificently elegant manifestation of projection, design principles, de facto, which exist, since the initiation of sequence, and certainly prior to this profoundly important to our species fraction of sequential event.

Ancient Masterpiece

Brilliant ancient art, de facto, first human artist, paint on the walls of caves masterpieces, yet, the composition of the masterpiece was already aligned with the progression of Intelligent Design, since the Initiation of Sequence.

Within the fabric of man is embedded creative power of composition, inherited by the grand plan of progression toward higher order. From mimicking and copying the reality, man is navigating to the very essence of transformation, creation of its own reality, based on proven definitions, with regard to the molecular, as well as ethical code … .

**Progression of Intelligent Design is perpetual. Molecules, energies, sequence
of transformation is designed to maximize the potential, human potential, de facto.
Good plan, indeed … .**

The Paramount Law of Transformation
vs
Perpetual Travel

Expansion of Parallel Universes
Sequential Travel vs DNA

The Paramount Law of Transformation
vs
Perpetual Travel

Univ I

Univ II

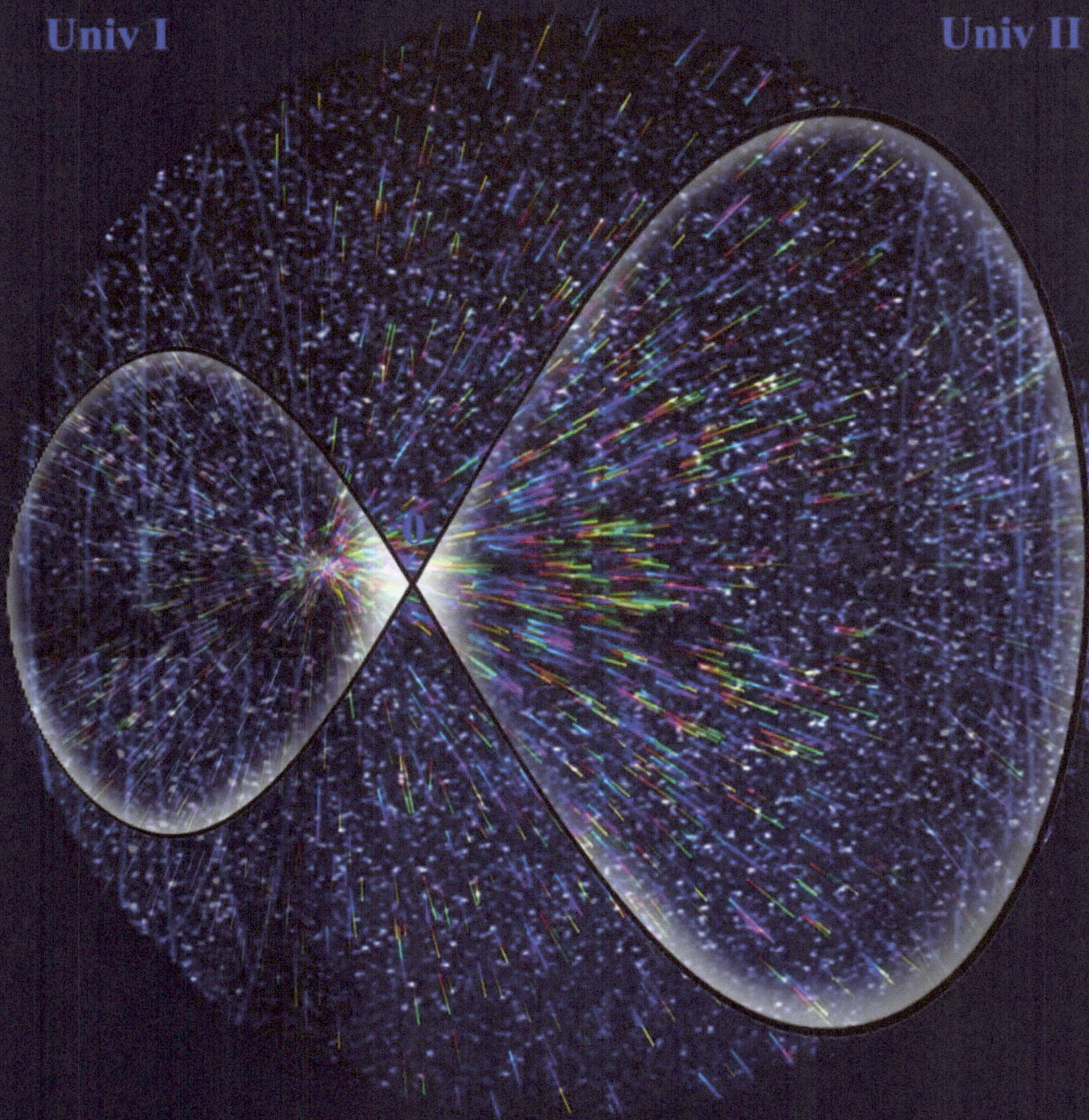

The Paramount Law of Transformation
vs
Perpetual Travel

Biological Blueprint of the Universe is compatible with human DNA, which, through progression of molecular, as well as energetic projections, allows for sequential traveling (time travel). As I have already indicated, time is irrelevant on Universal scale, yet, sequence is accessible at any partition of development.

The notion of perpetual existence is multidimensional, where sequence is progressing beyond boundaries, as well as limitless accessibility with regard to sequence within. De facto, so called time travel, is not only possible, but allows for accessing, perpetually sequential partition.

Biological Blueprint of the Universe, DNA structure in particular, represent potent model with regard to sequential exploration, structural molecular blueprint, and most of all, sequential travel, de facto.

DNA is progressing, yet, despite social molecular parallels and complexities, is also projecting profound perpetual potency, simplicity in complexity, and progression toward higher order of existence. DNA allows to explore fragmentation of sequence within biological existence, as well as progression of existence on Universal scale.

Projection and expansion of parallel universes is, de facto, possible through the implementation of the DNA model, from initiation of suquence, similar with each human, as well as entire human race is progressing.

Due to the progression of universal sequence, past and the future is interconnected in a seamless fragmentation of projection, and this structure allows for changing location, and is the essence of perpetual existence within non perpetual sequence.

Perhaps, we live in already non existent world, which progressed beyond the status quo of progression, yet, by accessing past sequence, past and the future world is „stretching" beyond physical and energetic limitations.

Fragmentation of seamless universal progression, Biological Blueprint of the Universe, DNA, allows for „stretching" of Universal projection infinitely ... .

The Paramount Law of Transformation
Expansion of Parallel Universes
Sequential Travel vs DNA

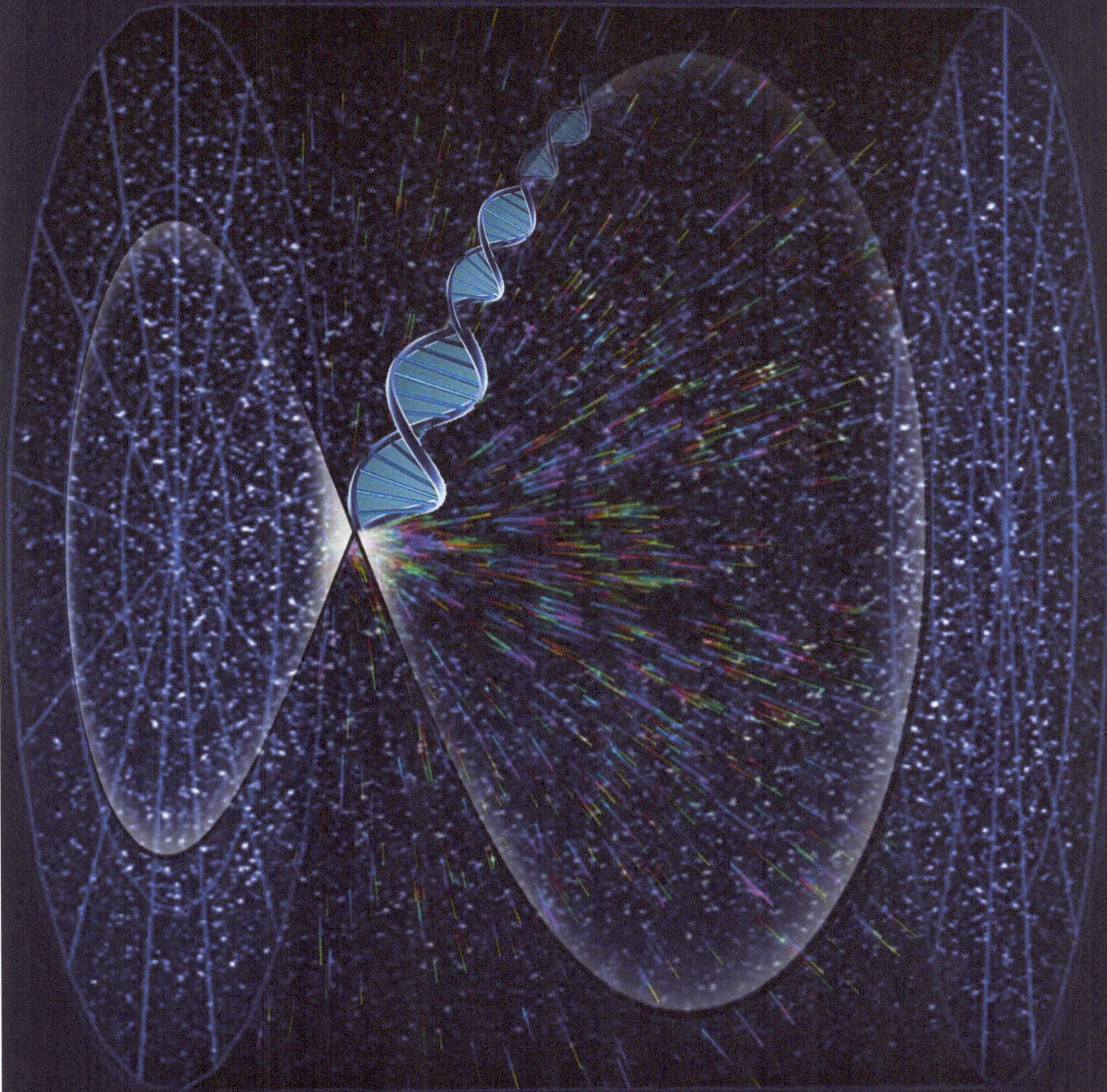

The Paramount Law of Transformation
The Essence of Intelligent Design

The Paramount Law of Transformation is projecting essential quality with regard to proportions, symmetry, de facto Intelligent Design, since the Initiation of Universal Sequence.

Universe, molecules, energies, are predominantly symmetrical with regard to structure, interactions as well.

Universe is abundant, de facto, yet, not excessive with any integral fragmentation sequence, beyond necessity of fulfillment. **Intelligent Design** represent engineering of common sense, where aesthetic component play defining role with regard to manifestation of progression. **Symmetry** is profoundly projected in biological data of progression.

Universe is speaking eloquently about symmetry, as well as aesthetic paradygm projected in **Biological Blueprint of the Universe**, as we know it, where geometric vocabulary is, de facto, parallel with algebraic eloquence.

If mathematics is unable to project geometrical form, than mathematical formula is either incomplete or false. Yet, at the same time, every geometrical projection is easily and eloquently translated into the mathematical vocabulary. **Universal Paradygm** with regard to Universal progression, as we know it, project automation projection, yet, with recycling system, where chaos and waste is non existent.

Intelligent Design, within **The Paramount Law of Transformation**, manifest **3D Projection,** brilliantly composed data, based on two compatible opposites, geometry and algebra.

3D geometrical projection is exceptionally appealing by the definition, yet, the mathematical vocabulary, represent practical application of geometrical vision, an idea. Once geometry and logarithmic equivalent blend as compatible opposites, than the foundation of projection, **Intelligent Design,** is formulated into the sequential data of complete **Universal Projection.** Why geometry and mathematics, are defined as compatible opposites data ? Because both project complete design. Geometry and logarithmic equivalent, manifest both, theoretical, as well as practical design.

I would like to also emphasize importance of sensual and emotional aspect of physical, molecular progression. All laws, as well as the entire variety of molecular components, are extended into the sensual projection. There are no separate set of Universal tools and laws. Universe, throughout, emphasize unification of all tools, laws, according to the **The Paramount Law of Transformation,** including Biological Blueprint of the Universe.

The Paramount Law of Transformation
Intelligent Design vs Design Principles
Geometry & Logarithmic Sequence vs Compatible Opposites

Golden Ratio

$$\varphi = \frac{1 + \sqrt{5}}{2} = 1.6180339887\ldots$$

Nautilus & Spiral
Progression of Fractals

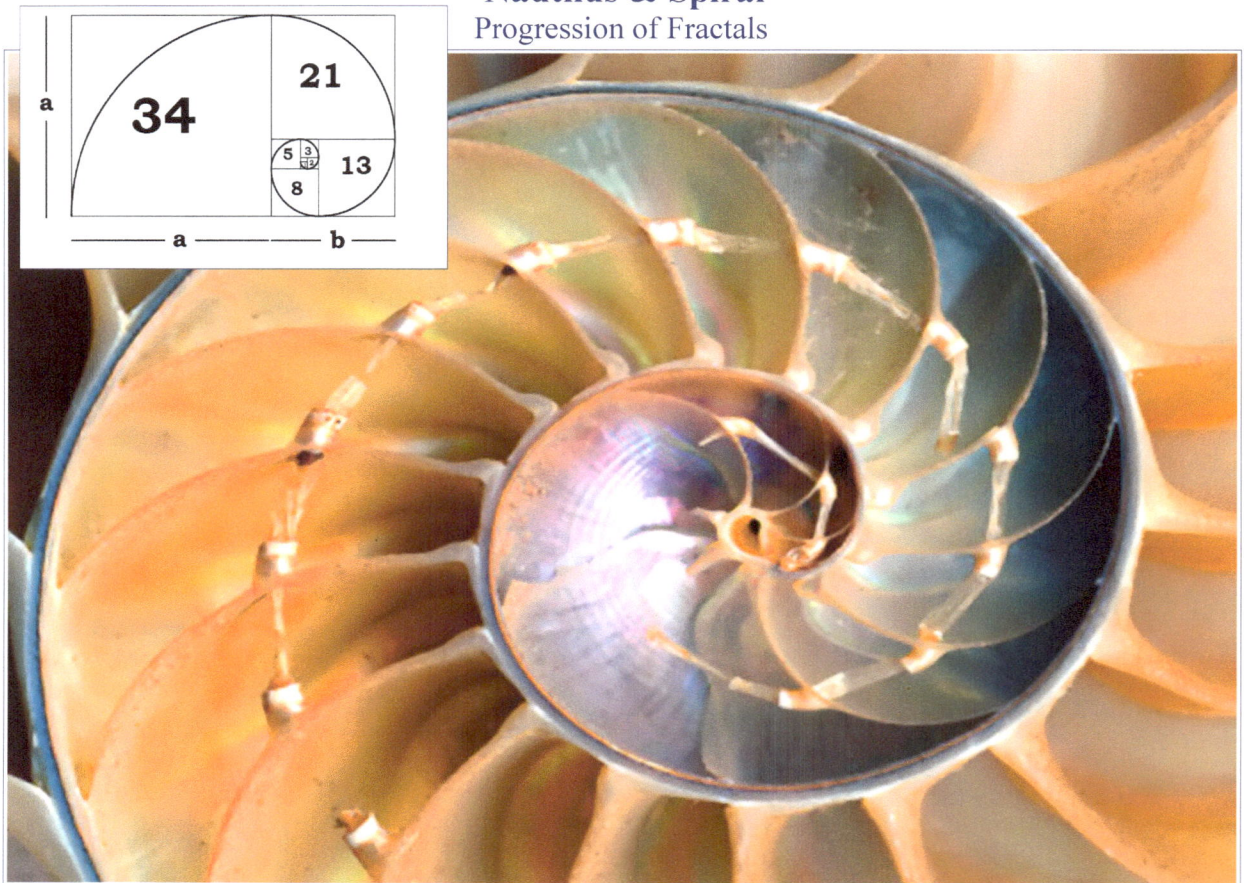

Golden Ratio represent, de facto, initial blueprint with regard to the complete design, where color variations, functionality, molecular, as well as central management system (nervous system in biological projection), physical components, location with regard to an application, are embedded into the seamless, beyond fractals, design sequence, as an attached illustration of Nautilus shows.

Above example illustrate design principles, as well as Geometry and Logarithmic sequence within implementation of compatible opposites.

The Paramount Law of Transformation
vs
Fractals Progression

The Paramount Law of Transformation

Intelligent Design

Progression of 3D Fractals
Nautilus, Fibonacci Spiral
resulted in projection
of the Sphere

Sphere

Spiral Progression
Sphere

Geometry
&
Mathematics
Compatible Opposites

The Paramount Law of Transformation
Geometry & Logarithmic Equation

The Paramount Law of Transformation
Intelligent Design

Water Molecules   vs   Ice Molecules

Free, liquified molecular automation.   Solid as well as crystalized automation: strictly defined molecular. Projection.

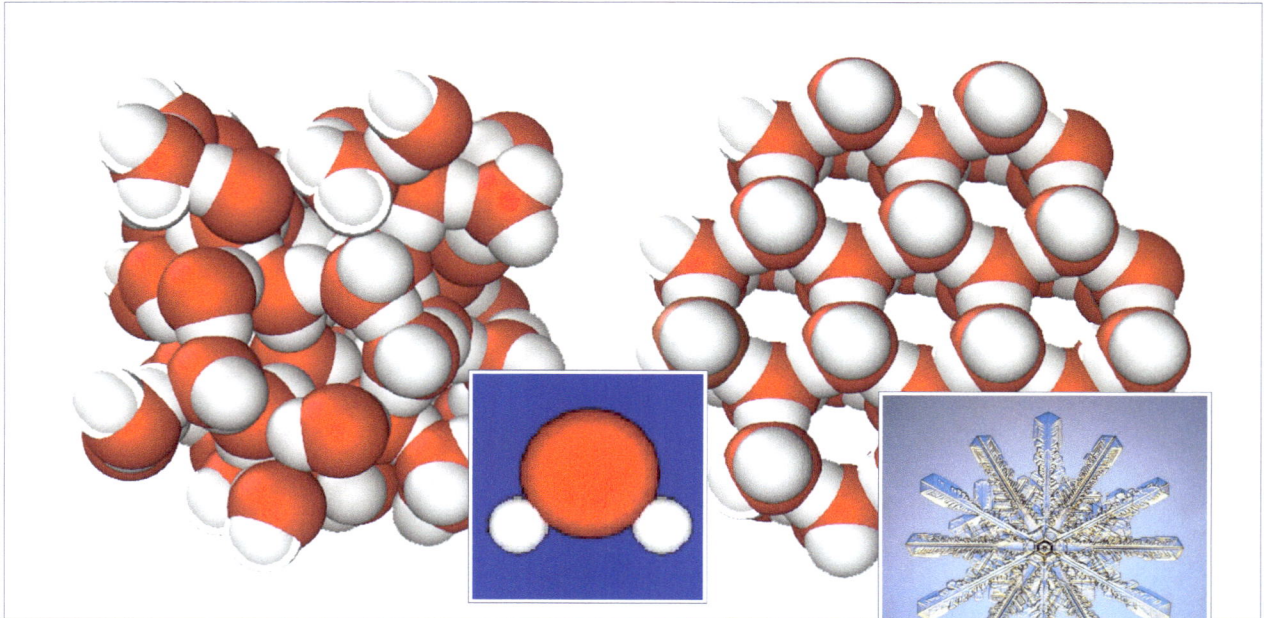

(Wikipedia picture: :non commercial use)

vs
The Paramount Law of Transformation
Progression of Fractals
Parallel Compatible Sequence
Golden Ratio vs Water and Ice Molecules vs Simplicity of Complexity

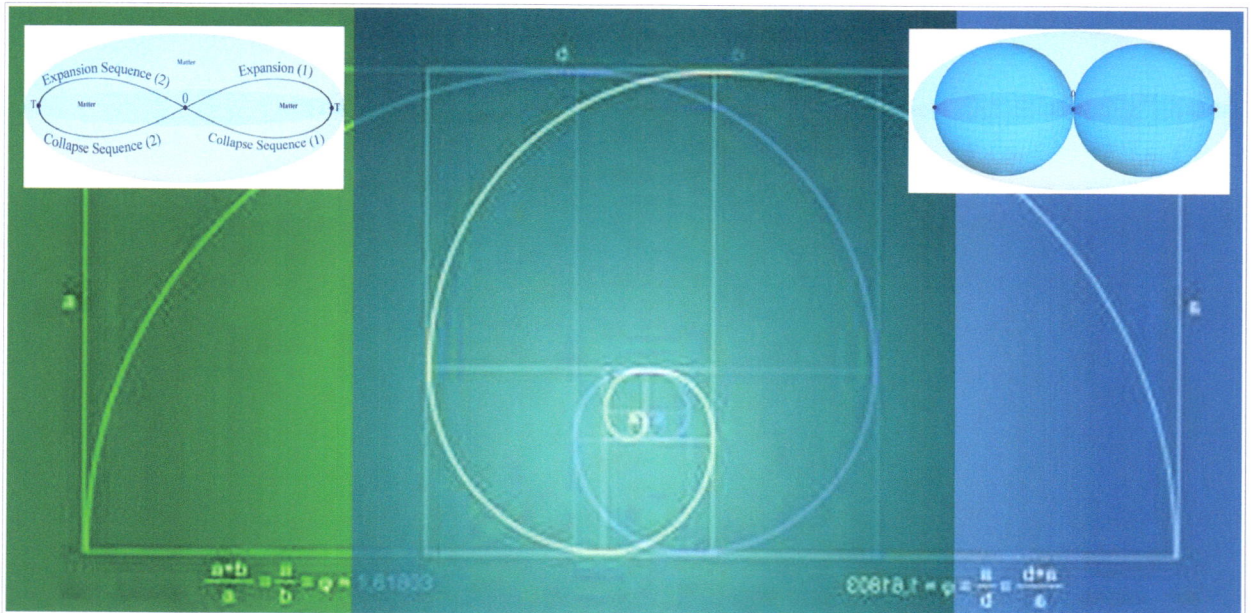

110

The Paramount Law of Transformation

Molecular Progression vs Aesthetical Projections
Aesthetical Universe vs Intelligent Design
Sequential 3D Progression Toward Higher Order

Intelligent Design

Flower: molecular projection, composition of data. Progression of Sequence: appr. 13.75 billion years.

Intelligent Design, data, is reflecting its own composition of data, properties: progression toward higher order, aesthetic paradygm within Universe, 3D projection, varieties within its own kind, simplicity, molecular automation beyond repetition of patterns.

Universe is progressing toward higher order, with regard to sophistication of projection, subsequently biological, organic forms of existence emerged, including sensual, as well as composition of aware energy.

The Paramount Law of Transformation
Biological Blueprint of the Universe vs Water. Hot Light vs Cold Light.

Human is composed with up 75 % of water (statistical). Apparently, we have the reference (approximation) with regard to the Universal Intelligent Design composition, proportions of Hot Light vs Cold Light. Water became miraculous, essential element of biological existence, along with light and other elements. Molecular simplicity in complexity is evident, progression toward higher order, 3D projection, Intelligent Design, de facto, is eloquently illustrated by the progression of sequence, since the universal initiation.

Human manifest, de facto, Biological Blueprint of the Universe, as we know it. Embedded DNA data, physical principles, elements and energies project efficiency, and at the same time, aesthetic beauty of data, in every composition, fragmentation of sequence.

I believe in Intelligent Design, brilliantly Composed Universe ... .

The Paramount Law of Transformation
Universal Progression of Compatible Opposites

Complete History of the Universe
vs
Biological Blueprint of the Universe

Energy + Inanimate Matter = Living Matter
(Hot Light + Cold Light = Initiation of Perpetual Universal Progression)

Initiation of the Universal Sequence through Compatible Opposites

$$1^\wedge + 1 = \infty$$

Transition of Living Matter into Aware Matter and subsequently into the Aware Energy.

The Paramount Law of Transformation
Universal Progression of Compatible Opposites

Complete History of the Universe vs Biological Blueprint of the Universe

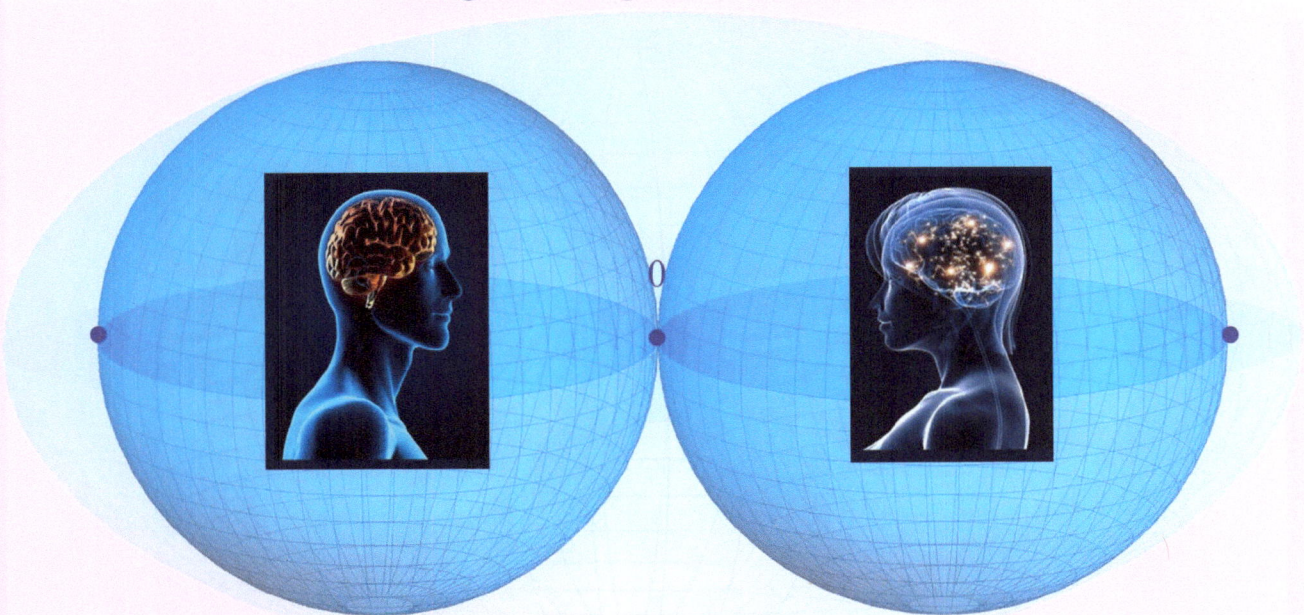

$$1\wedge + 1 = \infty$$

Energy + Inanimate Matter = Living Matter
(Hot Light + Cold Light = Initiation of Perpetual Universal Progression)

Initiation of the Universal Sequence through Compatible Opposites

Transition of Living Matter into Aware Matter and subsequently
into the Aware Energy.

Biological Blueprint of the Universe

The Paramount Law of Transformation

Computation of Matter vs Progression of Biological Molecular Sequence
Biological Progression vs Energy of the Future
Progression of Sequence vs Unified Progression toward Higher Order

Computation of Matter vs Progression of Biological Molecular Sequence

Computation of matter is a proven phenomenon, which occurs throughout the Universe, yet, often available visual simulation is limited to progression of the same elements into object, which is cloned from the original data. In Universe as we know it, computation of matter and subsequently energy, is based on progression of elements, yet, developing higher order in all directions. This is the essence of computation of data in all known DNA systems, either within a living matter, unaware, and profoundly manifested in progression of living, aware matter.

Progression of matter, developed biological data, as well as sequence of progression. Biological existence is the direct result of data, which is predesigned to reach higher order. From biological sequence miraculously emerged aware biological manifestation of sequence.

Biological system, as well as sequential order is providing solutions for most, if not all, human essential needs, such as energy, including electricity.

Fauna and flora are producing pure, efficient and sophisticated energy source, which ought to be explored to satisfy essential needs of human civilization.

Biological electric farms can and will produce eco compatible energy source, in this instance, electricity supply.

Future of energy, biological electricity. Biological Farming.
(according to the higher moral standards)
Example.

Bioluminescence
Animalia Ctenophora. Comb Jelly. Greek: "com b-bearers".

Important disclosure; all matter, unaware as well as biological, shall be treated with care, and respect. Entire Universal system manifest unified data, living organism, profoundly interconnected, and require respect in all applications and implementations, yet, most of all, expanded frame of awareness, to grasp the notion that respecting nature, human manifest self respect, de facto.

Biological Coupling.
Progression form singular source into the complex biological systems.

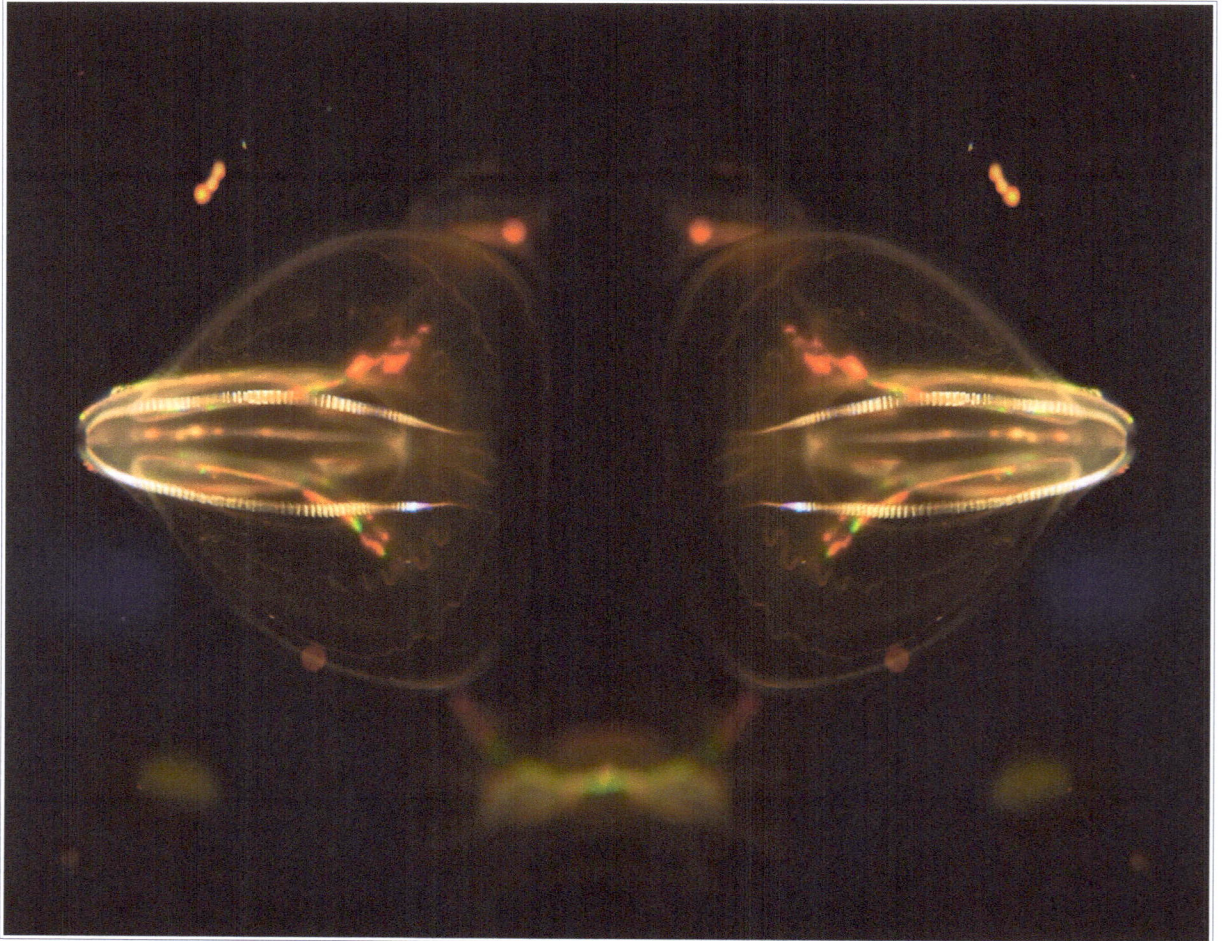

Bioluminescence
Animalia Ctenophora. Comb Jelly. Greek: "comb-bearers".

     Above illustration provide reference with regard to the possibilities, very much common in the Universe, as we know it, predesigned data, de facto, possibilities toward higher order.

     Nature provide endless solutions, as well as universal blueprint, computation of data, profoundly logical, sensual, and aesthetically appealing. Biological electricity = interaction of energy with matter. Biological solution to heal the planet, and satisfy essential needs of civilization.

     The essence of The Paramount Law of Transformation; essential rules, which manage entire universal system, perfectly implemented simplicity of complexity, unparalleled beauty of logic and grace … .

I believe in Intelligent Design … ..

Natural Biological Electricity. Sequence of progression from singular to complex.

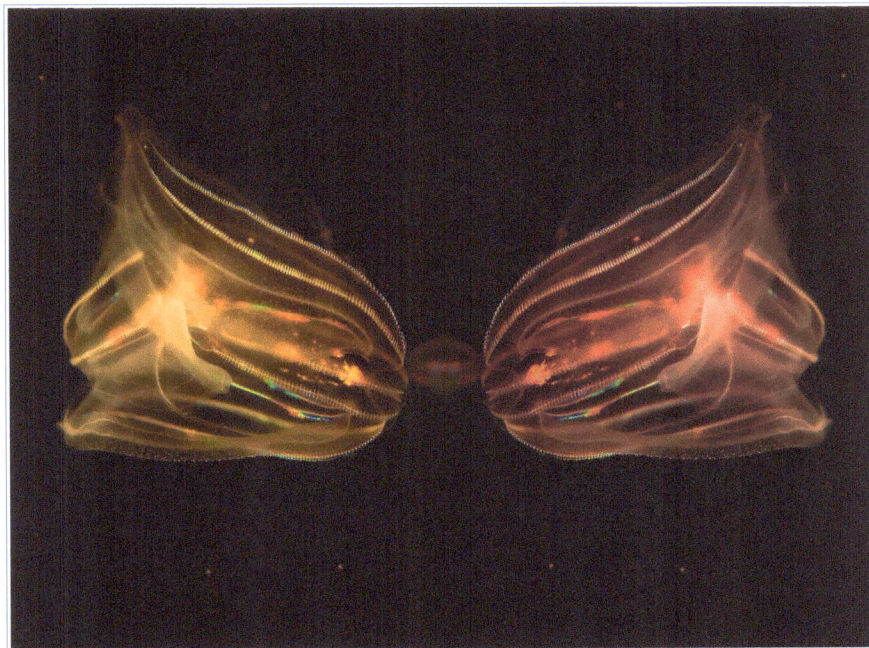

The Paramount Law of Transformation

Perception vs Past vs Present vs Future

Frame of awareness, as well as frame of perception represent physical, biological, de facto, occurences within interactions between molecular and energetic progression, subsequently the beginning of modern perception is dated to the universal initiation of data, yet, even prior to this phenomenon.

Sequence is a progression toward new horizon, yet, universal horizon is an illusion, due to the perpetual property of sequence.

Quote; „the difference between past, present and the future is just an illusion" (uncle Albert). He is right, above quote indicate that time is non existent, represent, de facto, elegant accessory of our civlization.

3D Cube is in perpetual motion, at any given moment. According to **The Paramount Law of Transformation**, motion is essential property of perception, de facto, through implementation of data, internal motion, external from subatomic molecular, to the enormous in scale galactic fragmentations of space. In this instance, motion occurs through interaction between compatible opposites, hot light vs cold light.

Perception. Rendering unified sequence toward higher order.

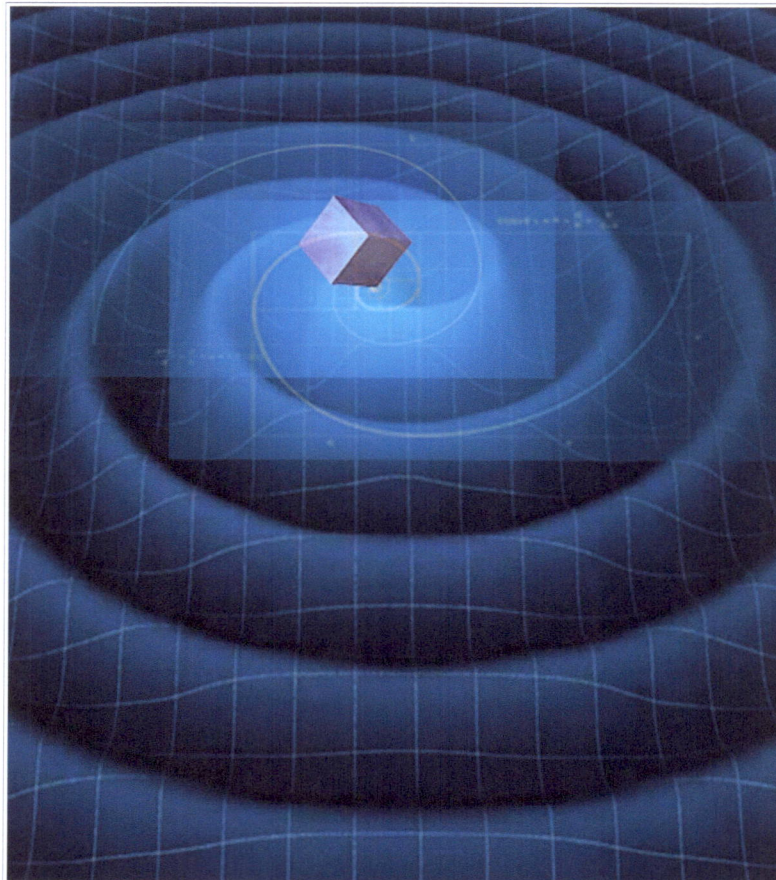

The Paramount Law of Transformation

Universal Progression vs Design vs Music
3D vs Flat

Music represent projection of higher order, with regard to matter, as well as energy, music manifest embedded data within DNA, inner desire, toward expansion of progression, frame of awareness.

How music is composed. Can we compose music without hearing the sound in our mind ? Certainly it would be impossible to compose a masterpiece without 3D progression of sounds within human mind. Sound is geometrical, de facto, **progression of electromagnetic energy,** energy, which is not only projecting 3D image, powerful force profoundly creating reality, reality virtually composed by the progression of molecular and energetic sequence, since the initiation of universal sequence, toward aware energy.

Sound manifest 3D phenomenon rendered through the flow of energy, and represent geometrical parallel with regard to mathematics.

Musical composition, classical masterpiece, by implementation of simplicity through complexity, represent fantastic model, simulation into the formation of the Universe, as we know it, and certainly beyond it.

Composition, in the beginning, is initiated by the inner sense of beauty, embedded aesthetic sphere. Idea is borne, idea progressing through (trial and error) the stages of composition, essentially 3D progression of aware energy within human mind.

The essence of composition (sequence of masterpiece):
Music (sound) = 3D geometrical rendition of an idea.
Composition, 3D sound (geometrical) = translated into the musical notes (mathematical translation).

Above represent complete, proper alignment with regard to the sequence of progression on Universal scale, yet, exact sequential paradigm of the Universe, Intelligent Universe, as we know it, where human project, de facto **Biological Blueprint of the Universe**, according to **The Paramount Law of Transformation.**

Idea + 3D Sound + 3D Composition + Musical Notes = Masterpiece

Universe, as well as the entire universal sequence is, de facto, 3Dimensional.
The notion of so called „flat" is non existent in any dimension.
3D progression, the essence of molecular and aware energetic universal performance.

The power of molecular, energetic progression of data, sequence of creation, require space to perform, and subsequently is translated into the algebraic language, which doesn't require 3D rendition.

Ravel's masterpiece precisely illustrate Universal progression of matter and energies, according to **The Paramount Law of Transformation** (Bolero played forward and backward).

3Dimensional world require space to perform, indeed, subsequently „world was created with the one purpose, music", the facto, essential substance within higher universal progression, where aware energy mark a new expanded horizon toward perfection.

I believe in Intelligent Design … ..

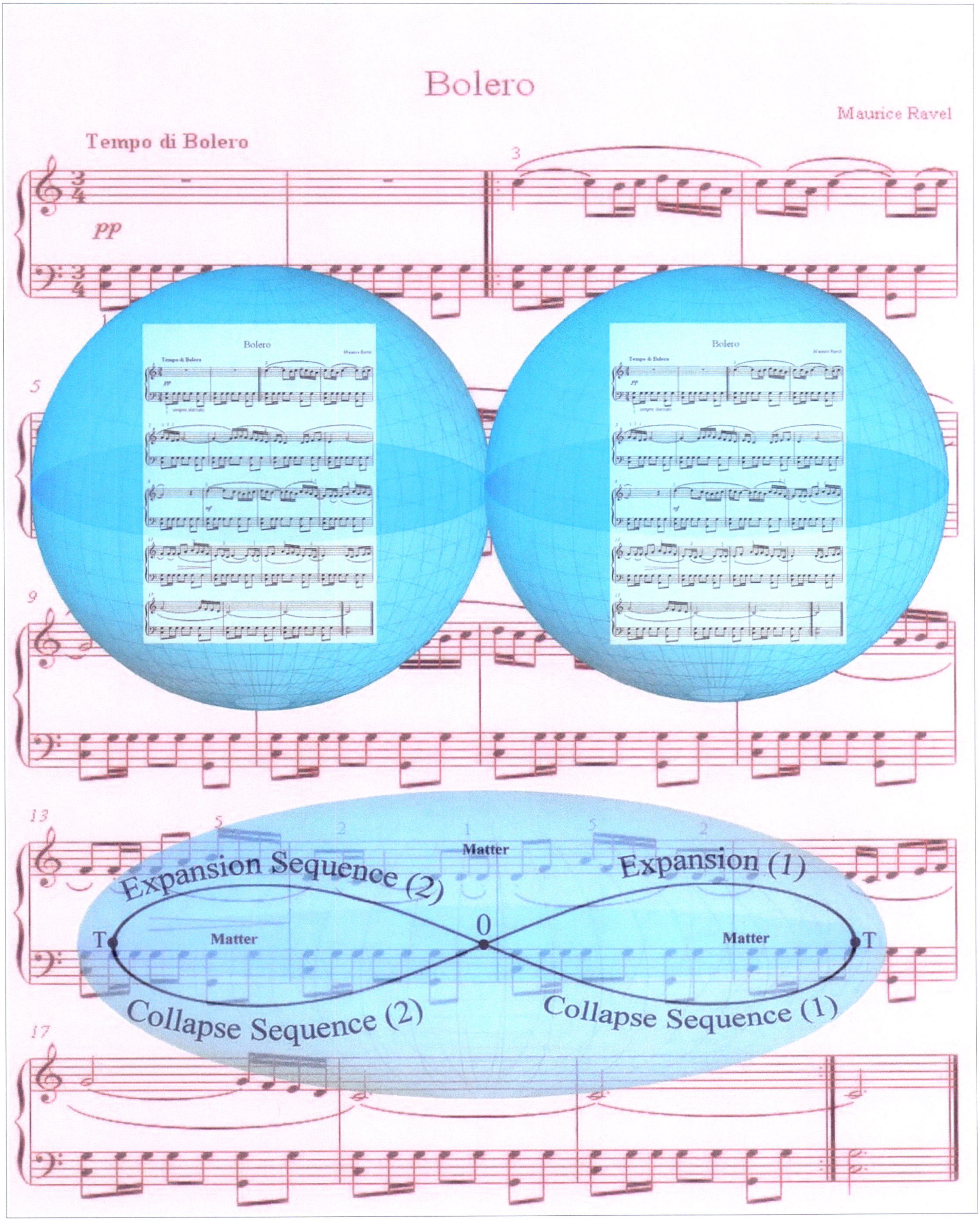

Bolero
Maurice Ravel
Tempo di Bolero
pp
Expansion Sequence (2)
Matter
Expansion (1)
T
0
Matter
Matter
T
Collapse Sequence (2)
Collapse Sequence (1)

The Paramount Law of Transformation
Big Bang vs Universe as We Know It
Universal Farming
Paradigm of Light

Universe is, de facto, a living organism, in every instance. Data of procreation and projection is navigating molecular embedded information, as well as energies toward higher order.

Big Bang vs Universe, as we know it. What we can assume, with a great degree of possibility, based on Biological Blueprint of the Universe, as well as Compatibility of Opposites.

Before Great Initiation Sequence took place, Universe existed, ready, potent. Primeval essence is dwelling (cold light), emotionless, yet, hibernating with miraculous potential.

Subsequently inanimate space is filled, fertilized with light, temperature. In an instant, Universal Initiation Sequence transforms, once again, space, which expands, and contracts like a vibrant heart, until data of transformation will reach a state, where energy of projection will progress into the subsequent stage of fertilization, other than Universe, as we know it.

Meantime, molecular progression reflect typical cellular, molecular replacement sequence, according to **The Paramount Law of Transformation, Biological Blueprint of the Universe,** which manifest a common, profound, essential quality, paradigm of illumination, which define projection vs inanimate state. Once again, **light is the essence of transformation,** despite sources of illumination, until progression of energy reaches a stage of aware energy, subsequently a stage of aware illumination.

Light, stubborn, truly miraculous phenomenon emerges through chain reaction, more or less subtle, sometimes violent, yet, never harmful with regard to the sequence of progression toward higher order.

Dark energy dwells, transforms, yet, it draws its potency from light, inverted, de facto, in physical, as well as philosophical, even ethical terms.

Paradigm of Universal Greatness is coupled with projection of existence, de facto, **Paradigm of Light** within molecular, as well as biological manifestations.

Every sequence is navigating toward illumination, and this is my friends, the **Genius of Intelligent Universe,** which transforms molecules, yet, aware existence, through exposing to the energy of light.

Is there any limit of the profoundly fertile energy of illumination ? Certainly not, if the essence of **Paradigm of Light** is creativity, in every scale … ..

The Paramount Law of Transformation
Energy vs Synthesis
Energy vs Data

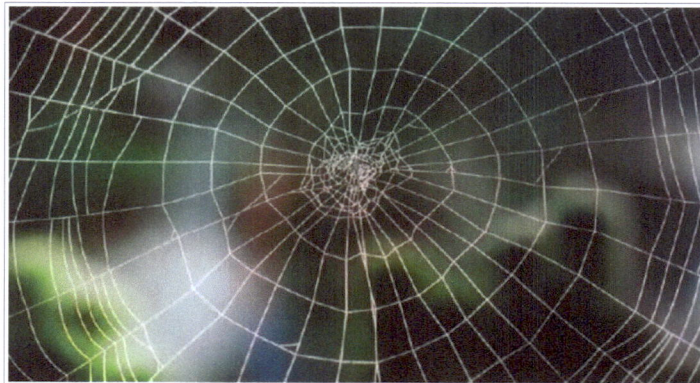

(Picture: Wikipedia)
Intelligent Design

All existent energy forms, within Universe, are compatible, according to The Paramaount Law of Transformation, compatible opposites. Energy vs data. Energy represents specific, unique data. Energy, every particle in natural world is either a real or potential form, source of energy. Yet, energy represent specifically formulated data, embedded into the molecular world. Entire system is based on few basic and de facto, simple principles: **data, molecular varieties, energy, de facto, transformation.**
Energy, as indicated in previous chapters, is an ability to transform, within specific anticipated design.

Yet, this process is progressing into the biological projection, and subsequently intelligent and aware energy.

Data of energy is transforming into the higher order, becomes, miraculously, destined to reach the very source of the Universal sequence, de facto existence. In this spectacular process is manifested unification between sequential fragmentation toward its source.

Progression of Unification Within Intelligent Design
Synthesis (all is one)

Higher Order of Intelligent Design = Data = Energy = Sequential Progression = Biological Projection = Aware Energy (data) = Intelligent Design = Higher Order

Human is, de facto „biological engine", which is formulating aware, intelligent energy, powered by light, water, air, as well as elements, which are shaping entire Universe, according to The Paramaount Law of Transformation.

Human invented, by the act of profoundly intelligent potent mind, analisis, great expression, synthesis, the very phenomenon, which is shaping universal data, including **unity, fragmentation, and sequence toward unity.** This is synthesis, de facto.

All that has been written in present, as well as previous articles, are provable, scientific discovery, based on Biological Blueprint of the Universe. Yet, discovery, perhaps do not reflect true nature of knowledge, which we ought to call, rediscovering Universal phenomenons, which were, and will still amaze our senses.

With regard to practical applications, a real or potential form, and source of energy, within vicinity, more or less remote, will allow, to obtain energy from anywhere, from saturated with light space, particles of cold and hot light, as well as every existent, in an instant. The engine of the future will use, absorb universal fuel, which translate as an existing particles, energies, yet, human will travel, ultimately, by the power of its aware energy.

Human is the ultimate source of inspiration, as well as solutions, on truly Universal scale, and certainly beyond it. I believe in Intelligent Design.

The Paramount Law of Transformation
Divine In Vivo Synthesis
(Latin:"within the living")

In Vivo precisely describe our home, Universe, which is, de facto, in all forms and manifestations, a potent, vibrantly living organism, aware and unaware, not limited to the molecular, yet, equally applicable higher order of biological projection, which emerged from a very molecular In Vivo progressions.

Yet, In Vivo carries another essential quality, **Universal Synthesis**, which by the definition, manifest data, program, an ability to transform by blending different molecular and energetic fragmentations, into variety of new projections. Synthesis is unique, yet, data, which allow synthesis to occur, represent comforting mystery of The Paramount Law of Transformation.

On Universal scale, temperature and light is the catalyst, which allows synthesis to occur. In Vivo synthesis is common within the Universe, yet, synthesis occurs, due to the embedded data, molecular formulation, where light, along with temperature is synthesizing world, within in vivo projections.

The science of **Divine In Vivo Synthesis** is profoundly mirrored in biological progression, according to **The Paramount Law of Transformation, Biological Blueprint of the Universe,** where process of unification, within compatible opposites is occurring, biological synthesis, catalyst during initiation process is also observable. Yet, all of that can happen due to the predesigned data within molecules, energies on every level of existence, where subsequently **Universal Potency** is equivalent with molecular world, energies, biological, biological aware, as well as, and perhaps the most, the **potency of aware energy**. From the Universal initiation sequence, up to the biological awareness, Universe is steadily increasing its creative potency, same as subsequent human generations.

Universe is evolving through Miraculous In Vivo Synthesis through embedded data of potency, which also increases, de facto, steadily.

Described phenomenons are not only observable, but proven through experimentation, scientific, up to the point, where you and I, are able to write, read, discuss above articles.

In Vivo Synthesis project its qualities as Universal data, de facto, **The Paramount Law of Transformation**, fragmented into the essence of compatible opposites. I would like to mention that all physical properties, in terms of molecular, energetic existence. manifest similar, if not identical psychological, social, emotional properties, within our own Divine Progression.

Universe is, de facto, one world, interconnected, undivided in terms of occurring phenomenons. What you see, and experience in distant galaxies, as well as sequential frame, is happening within our own vicinity, including inner, emotional, physiological, physical, internal, as well as external, social, biological.

Universe project unified sequence toward higher order, well designed, managed, maintained world, world, which is sharing exact the same properties, phenomenons.

Universe is a miraculous, living organism, driven by **Intelligent Design**, design, which human being is experiencing in every manifestation of existence, the facto, **Biological Blueprint of the Universe, Secret Code of Existence,** wrapped in the gift box, **The Paramount of Transformation**, with written code: enjoy life, divine gift, one of a kind … .

The Paramount Law of Transformation
Biological Blueprint of the Universe vs Perpetual Engines

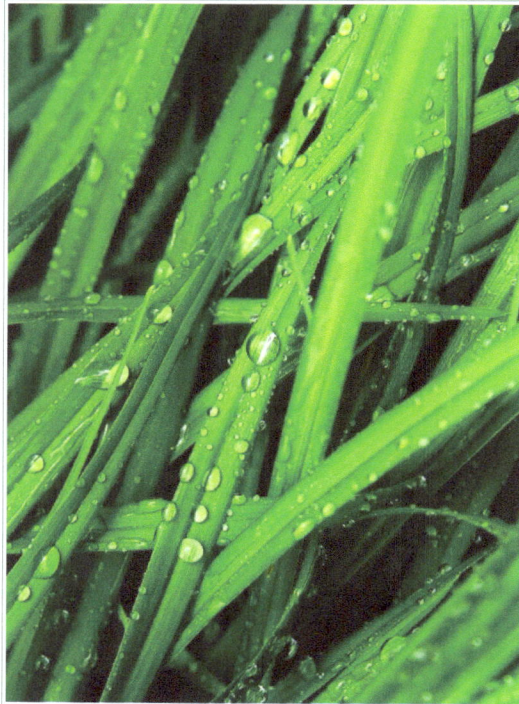

Human being is powered by various energy sources: air, light, water, and molecular compounds, such as vitamins, carbohydrates, sugars, fats, to name a few. Human biological engine is using variety of fuels, to support complex demands with regard to proper functionality, functionality, which resemble fundamental, and common quality throughout the universe, transformation toward higher order, sophistication, de facto, parallel system with regard to the Universal engine of transformation, where variety of materials, including hydrogen, oxygen, helium, coal, water, silicon, oxygen, and other elements, as well as energy sources, light in particular, along with associated spectrums of light, carriers of fundamental data, and cold light, ever existent background.

Human design strongly indicate that is based on perpetual propulsion ability, due to the Intelligent Design, which transforms and adapts to the variety of challenges, evolve to the higher order within and beyond each generation.

Human miraculous functionality indicate how to design engine, propulsion system, as well as adaptability of variety of energy sources, common within vicinity of the Solar System, as well as remote neighborhoods, yet, often similar in terms of mechanisms.

Another proposition of the Universal Design, also present within Biological Blueprint of the Universe, is gravitational propulsion, which hold everything within anticipated order, locally and globally, on Universal scale. Galaxies, clusters of galaxies, are powered by the gravitational engine, the very same engine which provide propulsion of the entire Universal system according to The Paramount Law of Transformation.

Gravity, the engine, is based on principle of compatible opposites, light vs cold light, two exceptionaly common and eloquent phisical phenomenons.

Gravitational engine not only provide propulsion, but transforms elements, and even energies, the very same system existent in Biological Blueprint, human eye for example.

Photosynthesis can serve in variety applications, where energy of light becomes an energy source , along with production of oxygen. Perfect solution for propolsion, as well as sustaining biological life, oxygen as well as source of energy (food) even source of electricity. Photosynthesis is truly divine solution for many applications.

The Paramount Law of Transformation
Biological Blueprint of the Universe
Positive Pressure vs Negative Pressure vs Space
Brain vs Progression of the Universe

Biological Blueprint of the Universe
Positive Pressure vs Negative Pressure

Our common journey to find a reason, as well as empirical essence in terms of awareness, with regard to modern man, curiosity about human origin, purpose of life, subsequently kinetics of transformation, represent tangible progression of magnificent awareness.

People argue from all social spectrums, scientific, spiritual, personal preferences, yet, my personal journey is to find scientific, tangible provable arguments to the discussion.

Universe as we know it, is managed by a few essential laws within all living matter, living unaware, living aware, as well as living aware energy.

The Paramount of Transformation through **Intelligent Design** is propelling the Universe, as we know it and apparently projection of **horizontal** and **vertical** progression.

Positive pressure (+) and **negative pressure (-)**, which defy gravity within molecular biological world, illustrate The Paramount Law of Transformation, Intelligent Design **eloquent composition.**

Plants, trees, tall in particular, transport vast amount of energy, liquids, nutrients from root system up to the top, often hundreds of meters high, defying gravity, yet, without mechanical system is implemented (pumps), except negative pressure, through tubes, perfectly sealed, perfectly composed, perfectly efficient.

This system, by any means is based on Intelligent Design according to **The Paramount of Transformation.** Negative pressure vs positive pressure, and gravity will perhaps become the future of transportation in various applications.

Apparently, positive, as well as negative pressure is also implemented in human functionality. Negative pressure is also verbalized in Universe.

Biological Bluperint of the Universe
Nature Defy Gravity. Biological Engineering.

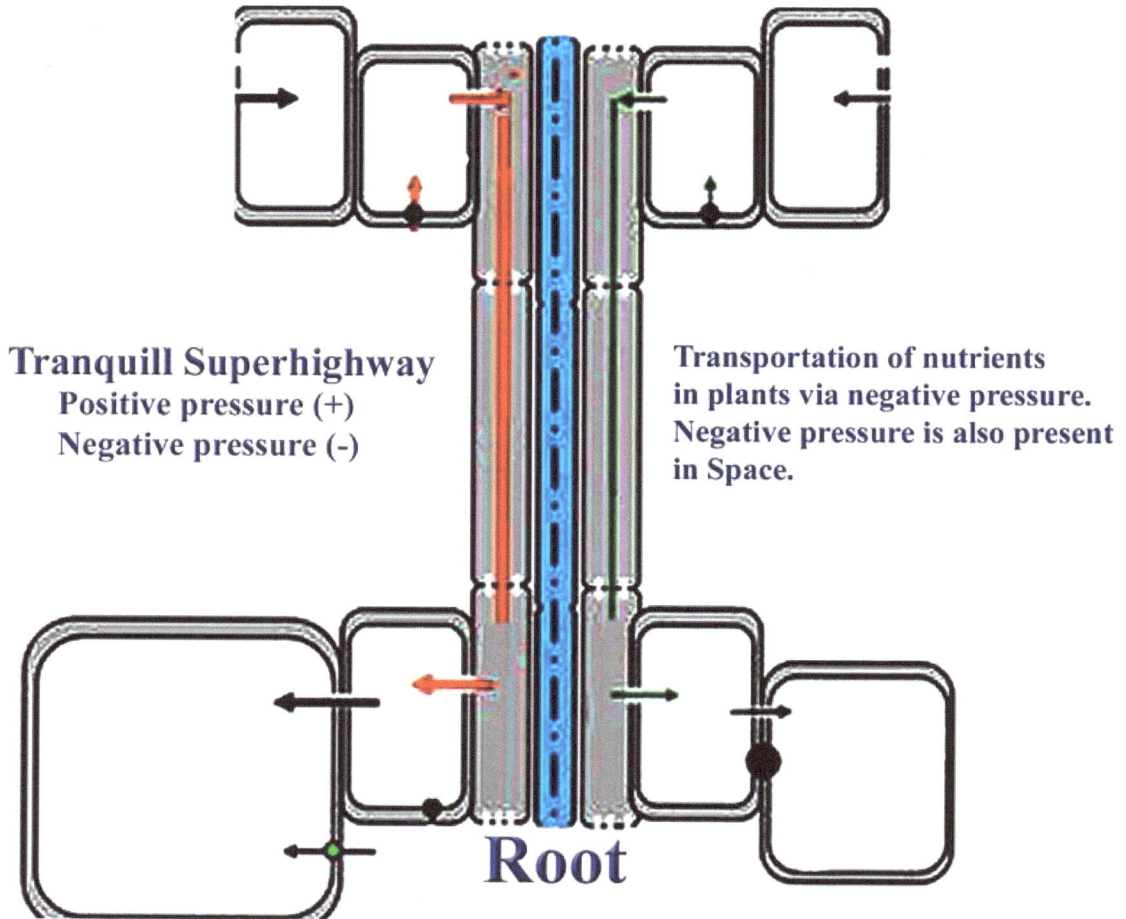

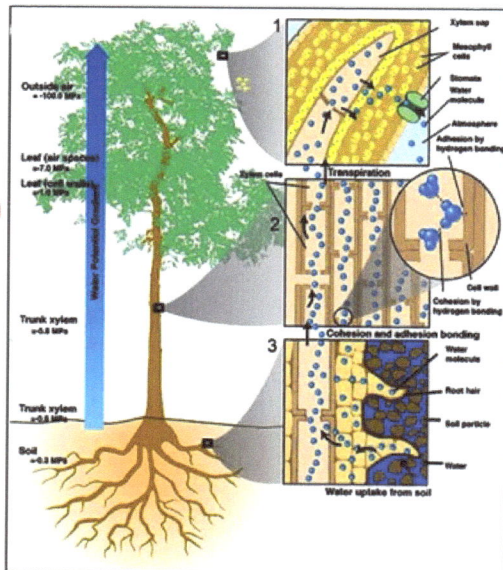

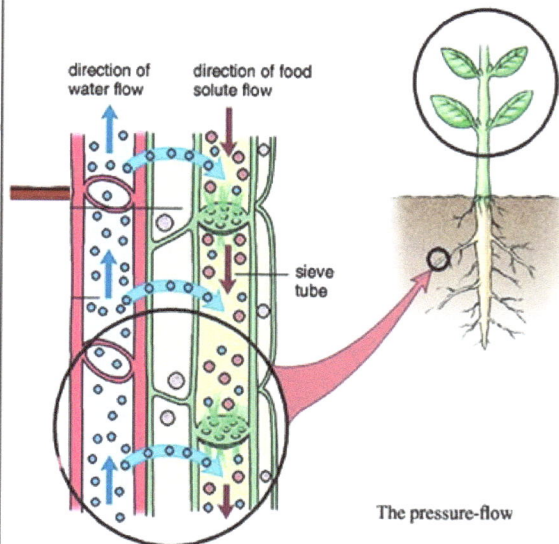

Tranquill Superhighway
Positive pressure (+)
Negative pressure (-)

Transportation of nutrients in plants via negative pressure. Negative pressure is also present in Space.

Root

The pressure-flow

Biological Blueprint of the Universe
Human Brain vs Progression of the Universe

Molecules do not think, yet, molecules are intelligent, due to the unique design, specific logical data, which dwell within each subatomic structure. Molecular DNA, by the definition, is equivalent with the paradigm of higher order, Intelligent Design, due to the specific information, written in each molecule, which becomes physical, as well as virtual carrier of progression, and apparently projection of higher order.

Human brain illustrate important parallels, relation between gray vs white matter vs Universe, composed by compatible opposites, hot light and cold light (light vs dark matter).

Universal architecture project its simplicity in complexity with eloquence shaped by **Intelligent Design,** de facto, **Paramount Law of Transformation.**

Scientific data is indicating, that relation between gray matter vs white matter is about 40 – 60 respectively. Yet, this is changing with every progressing generation.

Scientists argue that human brain varies in terms of efficiency, yet, one indication is constant, human brain does expand in terms of its potential, same as the Universe. In addition, molecular, physical, practical functionality is decorated by unique vocabulary of Intelligent Design, aesthetic paradigm, not as an accessory, but the essence of **Universal Design**, I have to mention once again. This feature of Intelligent Design is very appealing to me.

Functionality of the design is embedded, interwoven with aesthetic architectural propagation, driven toward higher order, in human vocabulary, higher standards. Parallels are obvious. Human is reflecting the very essence of its own brilliant, divine design, as well as origin.

Once you read papers about Universe, as well as human brain, quote: "brain filled exclusively with nerve fibers appear as **light-colored** white matter, in contrast to the **darker-colored** grey matter, that marks areas with high **densities** of neuron cell bodies... ", realization about parallels, with regard to design, projection of reality are obvious, provable, attainable for experimentation.

Subsequently we can formulate following question: de we, humans, project **Universal Reality,** or perhaps **Universal Projection** via **Intelligent Design** manifested human awareness. De facto, **Human Biological Blueprint of the Universe** incorporate both. What a brilliant design this is, indeed.

NASA, quote: "It turns out that roughly 68% of the Universe is dark energy. Dark matter makes up about 27%. The rest, everything on Earth, everything ever observed, with all of our instruments, adds up to less than 5% of the Universe, it is such a small fraction of the Universe". Parallels between NASA, as well as paper about human brain (quoted above), is very interesting, indeed.

Human intelligence, brain capacity represent another platform for the discussion. I believe, that human brain, used today, varies from person to person, yet, it's greatly mismanaged in terms of objective, to increase its progression with regard to efficiency. Please also note that individual intelligence represent a fraction of the entire equation, because, we as a species, ought to recognize and focus on cumulative intelligence, as well as collective intelligence.

I am talking about natural brain capacity, possibilities, not a technological experimentations, and God forbid, implants. In this instance, so called 10% human efficiency, is a very generous estimate. Perhaps we ought to decrease efficiency, and increase its potential at this point.

Human brain, through Intelligent Design via molecular progression, represent unparalleled sophistication, with regard to molecular unaware existence, molecular aware biological existence, up to the point of aware energy. Technology, in one instance is propelling sophistication, yet, is decreasing brain capacity to progress beyond feasible horizons, yet, feasibility is a factor limiting instead of stimulating toward unlimited potential, progression, de facto.

Humanity ought to stimulate its awareness toward natural brain development, the venue of true progress, which technology will perhaps never achieve, yet, if it will, that would be undesirable scenario for human race.

Educate, progress, manifest what you've got the best, unlimited self, along with ethical transference of inner, as well as external reality.

Universe, via **Intelligent Design**, is carrying humanity on its shoulders, let's learn from it, by carrying each other, through implementation of ethical behaviour, the very same behavior entire universe is progressing, progression toward higher order, de facto. This is the complete architecture of progress according to **The Paramount Law of Transformation.**

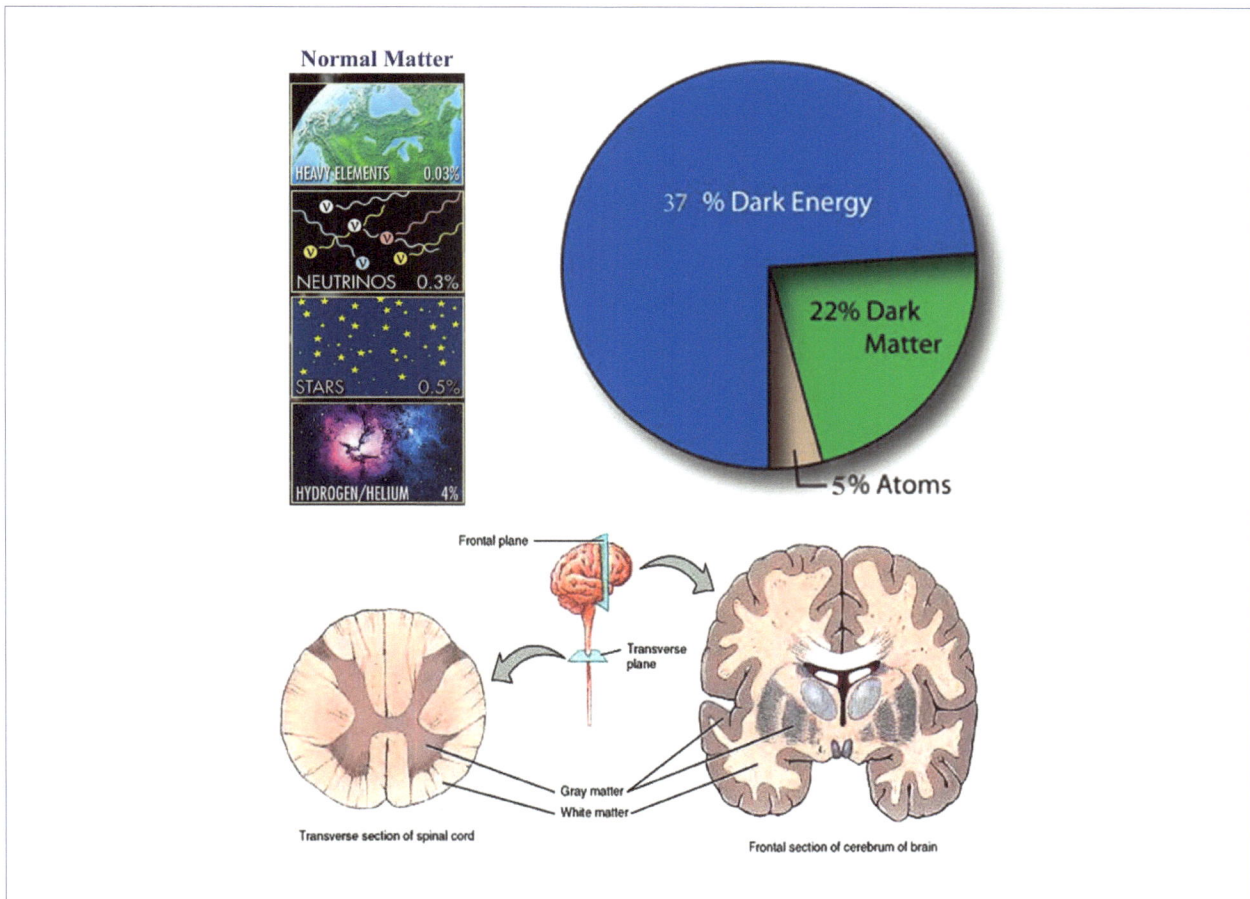

The Paramount Law of Transformation
Biological Blueprint of the Universe vs Resonant Echoes of the Worlds
Universal Resonant „String" Compatibility
Resonant Phenomenon

Human Brain Resonant Frequencies
vs
Resonant Frequencies of Parallel Universes

Resonance: Latin resonantia **"echo"**, from resonare, "resound".

Resonant Phenomenon

Resonance manifest phenomenon that occurs when a certain system; object, molecular, energetic projection, is initiated by another vibrating system or external force to oscillate with certain amplitude at a specific frequency. **Resonant Compatibility** is a key term, when we talk about resonance, because not all objects or physical projections, including molecular and energies will produce responsive vibrations along with resonance frequencies, where small periodic vibrating forces have an ability to produce large amplitude oscillations.

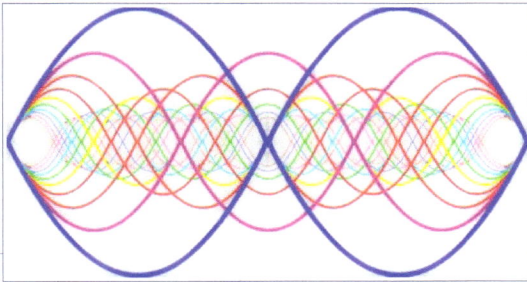

Multiverse Initiated Through Resonant Waves

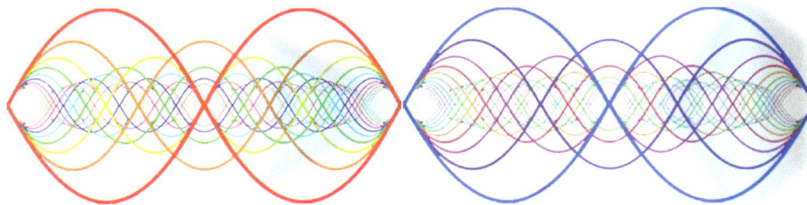

Resonant Phenomenon

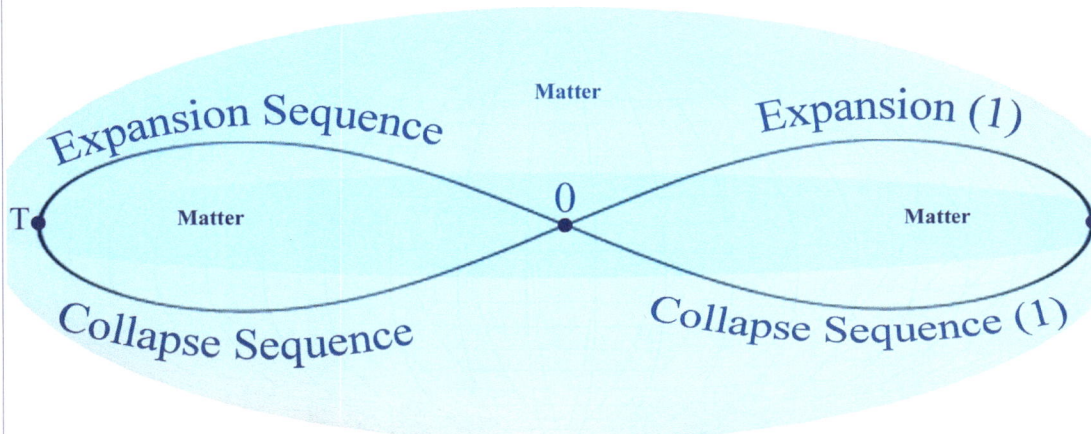

Expansion Sequence

Matter

Expansion (1)

T

Matter

0

Matter

Collapse Sequence

Collapse Sequence (1)

Resonance phenomenon represent, de facto, common universal occurrence equipped with different kind of vibrations, as well as waves: mechanical resonance, acoustic resonance, electromagnetic resonance, nuclear magnetic resonance, electron spin resonance, resonance of quantum wave functions, musical instruments, profoundly projecting its properties within **resonant frequencies of the human brain, s**ense of hearing. Light, de facto, electromagnetic radiation projected via resonance on an atomic scale, timekeeping mechanisms, tidal resonance, sound acoustic resonance, vocal resonance, electrical resonance (radio, television), laser optical resonance, orbital resonance within the solar system, electron spin resonance, nuclear magnetic resonance.

Universe, as we know it, is filled with compatible resonant "voices", where data of resonant phenomenon is incorporated within Universal reality. Data of resonant phenomenon is, de facto, the engine , which assist unaware, as well as aware existence, including aware energy toward miracle of transformation.

We could look even at the ancient texts searching for clues, for example Bible and other texts, where you can find, the very first and the most profoundly accentuated paradigm of existence, "first was the word", yet, in modern vocabulary, PhD physicist could say, it's about resonance and vibration, because it is, de facto, exact definition of electromagnetic resonance within the brain, as well as vocal cords. Archaic language is translated into the modern, demanding, sophisticated scientific language.

The Paramount Law of Transformation does perform because Universal resonant phenomenon is embedded into the physical properties, which progress beyond motion, an intelligent motion reflected through intelligent data from subatomic level. Resonant means, **"I am"**, it does represent, de facto, molecular language of interactions throughout the universe. What a great way to communicate, and translate data from one source to another. Resonant waves … .

And this how multiverse phenomenons are progressing through resonant interactions, parallel with human, social interactions.

Resonant Phenomenon is reaching beyond our own Universe, de facto, and is parallel with **Biological Blueprint of the Universe**, human interactions, indeed. Molecules, matter, energies, share common **Universal Quality, Social Property**, within DNA of the Intelligent Design.

Human manifest projection of social properties within matter and energies, so does Universe.

Resonant Phenomenon initiated Universal sequence, sequence which is reaching beyond boundaries of the Universe, as we know it, progressing into the propelling subsequent worlds according to **The Paramount Law of Transformation**. The very same system is profoundly present within human society.

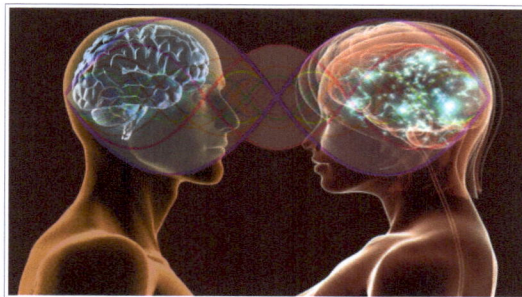

The Paramount Law of Transformation
Building Blocks vs Universal Hierarchy of the Design
Molecules in Multiple Locations

Universe wasn't build by building blocks, but profoundly potent intellectual process, Universal Laws, de facto, according to the **The Paramount Law of Transformation, Biological Blueprint of the Universe**, which indicate that Intelligent Design would incorporate variety of materials as well, including unknown to man, in order to build Universal sequence of hierarchy, with regard to the progression toward higher order, projection of existence, aware and unaware, as well as progression toward aware energy. To date, we recognize few dozens building blocks, which actively participate in The Paramount Law of Transformation grand program.

Universe, as we know it is, de facto, The Paramount Law of Transformation, defined by set of simple laws, **intelligent data**, which propel progression of sequence, which include, among others, **the law of compatible opposites**.

The Paramount Law of Transformation program, simplified blueprint of **Intelligent Design Hierarchy** :

* design of laws, data, embedded within intelligent universal sequence
* design of building blocks, which would participate in the program (from subatomic level)
* design of space, where sequence would perform
* design mirroring self replication projection data at the advanced sequential projection (biological)
* projection of progression of aware energy sequence.

Building blocks reflect virtual and physical potential, yet, Intelligent Design is, de facto, hidden from the obvious, external layer of awareness, which through principles of the design, intellectual potency, knowledge, creativity, design blocks for entertainment, education or other projects. The very same sequence took place before, during and after Universal initiation. Attached pictures illustrate this process eloquently.

Universe, as we know it, perform according to the **Intelligent Design** hierarchy, principles, de facto, laws, building blocks, space, mirroring, projection of higher order. Interestingly enough, ancient archaic language does mention about it, as well as modern science, which says, that we are the projection of Intelligent Design, which is progressing still, expanding horizons by incorporating intelligence and potency of possibilities, yet, the Universe can be replicated by the sequential pattern, past and the future (I've mentioned in the past, as well as previous chapters, that **present manifest appealing, elegant accessory of sequence**, yet, sequence is performing according to the principle of perpetual transitions, which varies in sequential intensity).

Building blocks:
* We can either design data of building blocks and influence the outcome, by the process of replication: we learn, de facto, how to design, based on proven scheme of logical possibilities.
* Building blocks may allow to incorporate unlimited potential of the mind, as well as necessary materials to perform, as mentioned previously (the future of the building blocks).
* We are somewhere in between the first and the second stage, profoundly important transitional stage, due to the fact, that human shall feel obligated to perform in advanced progression, higher order of progression, according to the moral laws, same as Universe does, which is holding civilization on powerful shoulders, even when the society manifest ethical tantrums, all too often.

Civilization is progressing, and demand for more intensity, in terms of implementation of ethical standards, which is being verbalized daily, provide promising development.

During technological revolution, building blocks, are designed by man and computers, which learn, at the same time, advanced universal principles of the design.

Images reflect compatible examples, leaf, fragrant oil, blocks for children, Iron Man, makeup set. Provided examples illustrate Intelligent Design, which is implemented according to the principles mentioned previously.

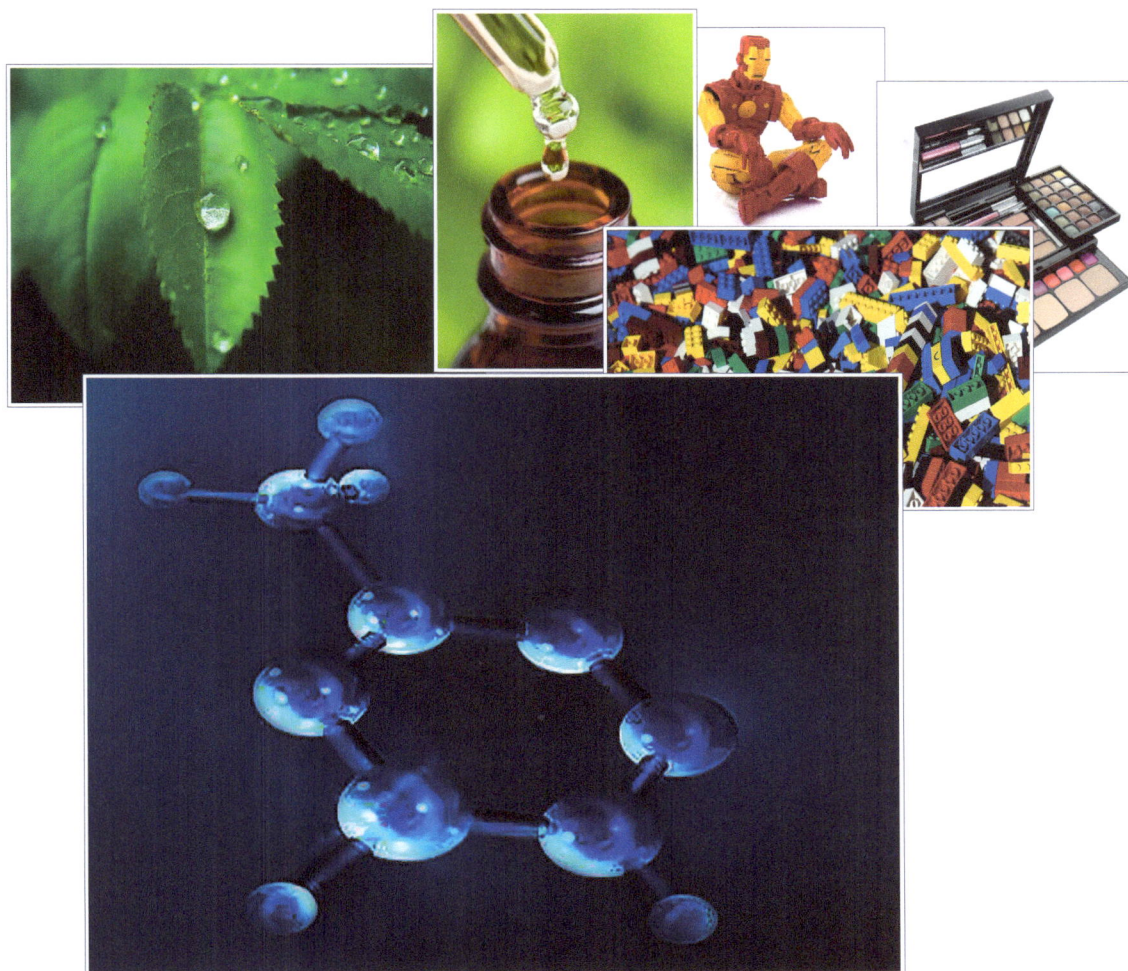

What we can learn. Everything happens according to the intellectual, empirical process, not by a chance, de facto, but through intelligent, advanced process. Behind each illustration is implemented unique design: building blocks, which allow children, as well as adults to entertain, expand imagination, and an ability to perform as an inventor. Makeup set formulated from a variety of materials, blends, combinations, fragrant oil, to enrich senses, extracted and prepared, mixed with other ingredients to achieve one of a kind scent for the purpose of aware, sensual decoration.

Molecules are designed to perform, socialize, transform, evolve toward higher order, miraculous sophistication according to **The Paramount Law of Transformation, Intelligent Design of the Universe.**

The phenomenon of bilocation, physically projecting in multiple locations simultaneously, is not only possible, but reflect sequential property of Universal projection, which by the very nature, is multilayered.

Classic example of multiple physical manifestations, are humans, who are „formulated" according to the same biological blueprint, molecular, particles and energies, yet, differ, same as molecules (every molecule is different, one of a kind).

When we look at another person or people, we realize that we see and experience (simplified example), de facto, the phenomenon of physical presence in multiple locations, performing within unique frame of sequence, yet, its own layer of sequential eloquence.

When I look at another person, or people, I see the phenomenon of bilocation quite common, (physical, molecular), yet, with its own spectrum of awareness, which is compatible with others. In the future, human will be able to experience the very same reality through another's person or people perception, yet, today we can simulate the process by intellectual, and even emotional implementation.

The Paramount Law of Transformation, **Biological Blueprint of the Universe** provide answers, direct references or fantastic shortcuts into the **Universal Design.**

Look into self for miraculous science, Universal Design, Projections … .

Once my article is published, than it can be present in multiple locations at the same time, and even projected within the mind.

Universe project unique vision, apparently Universe is performing via projections through progressions.

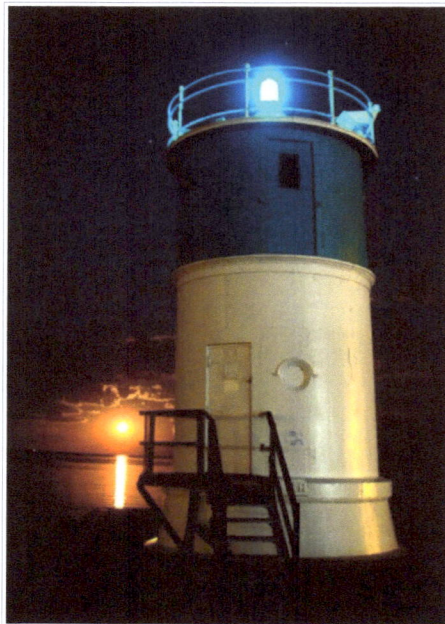

Reflections & Mirroring.
Universe is performing, Man as well … .

The Paramount Law of Transformation
Biological Blueprint of the Universe vs Universal Projections
Progressive Projections vs Reversed Progressions

Within the Universe, all physical objects, including human being will be able to reach technological stage, where **spontaneous transitions** from one location to another will be possible. Good example of disassembly and reassemble illustrate a water fall, water in general. From solid body of water, molecules are dispersing to once again reassemble into the solid, unified body of water. If simple molecules or molecular structures are potent with regard to spontaneous transitions, than transitions with regard to complex structures represent a real possibility.

Spontaneous projections are common within the Universe, as we know it, yet, complexity, for obvious reasons, is far more advanced and demanding. Ancient texts indicate exact phenomenon i simple yet eloquent phrases, stating, quote: „from ashes to ashes", which illustrate molecular transitional ability, from simple molecular structures into surprising complexities, and subsequently into simplicity, unless molecular and energetic progression reaches a stage of aware energy, than obvious limitations are no longer maintained. Human is composed from a variety of materials, yet, water is profoundly embedded into biological blueprint.

Molecular, biological structures manifest structural integrity (often maintained via powerful forces such as, in subatomic reality "strong field" or gravity), which progressed from simplicity with regard to design. If sequence is progressing toward complexity, than in sequential Universal progression, the process is reversible, and this is not a speculation, but a real, **common phenomenon** occurring within the Universe, de facto. Water, snowflake in particular, provide classic and precise example about the **Universal Possibilities**.

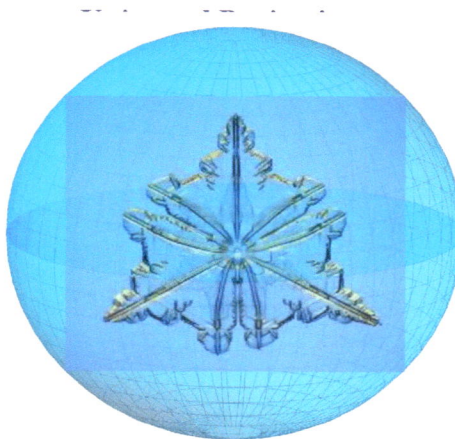

Progressive Projections   **Reversed Projections**

Universal Projections, graphically, are rendered according to the attached illustration. Universal laws progressively engage with practical solutions, yet, realization of transitional phenomenon manifest the most important step to achieve its potential and fertility.

Science shall serve the common good. **Universe, according to The Paramount Law of Transformation is, de facto, unlimited in terms of projections within progressions, yet, the only limitations are ethical.**

The Paramount Law of Transformation

Reality vs **External Reality** vs **Internal Reality** vs **Progressive Reality** vs **Projective Reality**

Reality, since the **great age of Greek** scientists, thinkers, philosophers, humanity is expressing intellectual, as well as vital engagement in this topic. What it is, how to define it, how to prove it.

Common scientific notion about external reality, internal reality, and so called "agreed reality" is important and require revision. Universe, in all miraculous projections, is progressing by implementation of simple paradigms, yet, humanity, often, perceive sophistication as complexity.

Universal Reality, proven and undeniable phenomenon is occurring within and beyond human perception, de facto. Human represent molecular, as well as energetic progression, since initiation sequence took place.

Universe, Intelligent Design is progressing toward optimal projection, same as human race. I would suggest to refrain from dividing reality as external and internal, because human perception, conscience project Universal Reality, which is, de facto, a „product" of progression, since initiation sequence took place, designed by the same elements, molecules and energies, as breathtaking Universe, yet, composed by divine projection, self awareness.

The destination of awareness is progressing toward aware energy, the source of Universal Progression.

Intelligent Design is progressing by maximizing its potential. Human, Biological Blueprint of the Universe, manifest this phenomenon with divine eloquence.

Universe is expanding toward higher sophistication, progressing via expansion of frame of projection, same as man does, de facto, through expansion of frame of awareness, according to **The Paramount Law of Transformation,** Biological Blueprint of the Universe.

The eloquence of the **Universal Reality** is projected via **molecular reality**, designed according to **The Paramount Law of Transformation**, to maximize its potential, as well as **biological awareness**, reality, translated, simply: „I am in you, you are in me".

Reality is manifested through projection of progression, simple, yet, beautiful paradigm, according to the Paramount Law of Transformation.

The Paramount Law of Transformation

Event Horizon Capsule
Vortex of Matter vs Data Preservation

Event Horizon Capsule
Vortex of Matter vs Data Preservation

Vortex of Matter (so called Black Hole) is transforming data, profoundly that is. Modern science is trying to grasp this phenomenon by formulating questions as well as simulating Vortex of Matter to understand perceptually, and practically **kinetics of Universal Vortex.**

Perhaps it's not as complex as it appears, yet, require extremelly advanced science to implement it. Vortex of Matter is created by compatible opposites. **Data preservation within Vortex of Matter** is possible and doable by creating equal vortex within vortex.

3D simulation of Event Horizon is indicating that by creating well balanced, precisely calculated **Vortex of Matter within Vortex of Matter** can, de facto, neutralize powerful forces and alloows travel through violent environment and ultimately provide, maintain data preservation.

Event Horizon Capsule sourrounded by its own Vortex of Matter, is capable to travel through Vortex of Matter, bumpy journey, indeed, due to the fluctuations within vortex of matter, yet, safe and efficient to overcome typical and existent physical obstacles.

The ultimate, yet, solution is energy, aware energy, that is, which is able to overcome any physical or molecular sequence of progression, because is the essence of sequential projection, the very fabric of Intelligent Design. The Event Horizon Capsule formulated by energy, is ultimate

Biological Blueprint of the Universe, according to **The Paramaount Law of Transformation**, is already performing simulation of vortex of matter in physical terms, everyday for a very long time, de facto.

Evergy does materialze, project not only reflected images, we call reality, but its own projections, as well capabilities to process variety of Universal projections, apparently human brain is performing these tasks effortlessly.

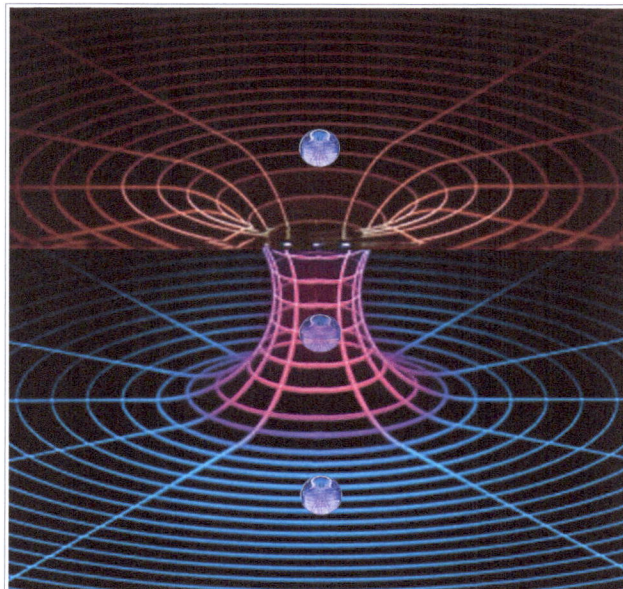

The Paramount Law of Transformation

Parallel Universes vs Cellular Automation.
Universe is within us. How to „Resurrect" the Universe.

The **Paramount Law of Transformation** is eloquently suggesting, based on Biological Blueprint of the Universe, that **Parallel Universes,** are progressing through cellular automation, preprogrammed to perform, according to the data embedded within. The process is parallel with human cellular automation as well.

Modern science is claiming that Universe, as we know, it is fading away, and apparently at some point will stop performing. Based on this statement of observation, I would suggest to look closely at the biological performance of human being to seek for clues and solutions. By the way, sequential performance, fragmented and global, is proving the very same principles, which are existent within **Human** as well as the **Universe** and **The Paramount Law of Transformation, Law of Everything,** which **unifies entire science** into the comprehensive set of awareness, laws, provable science.

It is true, that Universe will stop performing, according to the desirable progressions, which essentially are projecting the paradigm of existence, supporting aware, biological life. Same applies to human being, yet, physical properties, are progressing into the energetic projections. We need to find the a solution, how to interfere with the process of slowing down of progressively fading away Universe. Solution is within human, The Biological Blueprint of the Universe is providing solution in this instance as well.

Human represent beautifully composed progression, as well as projection of energy, electromagnetic, precise temperature, transitions, based on metabolism. What a great system, indeed. Nearly every part of human body is regenerating throughout life. Right now we know precisely where to look for the solution, fantastic development, indeed, but how to find it, is not as easy as it seems, yet, by maximizing any option to achieve desirable result, will produce a breakthrough.

Another important point I wish to make, is that entire Universal system is based on interconnected fragmentations, where human being is playing profound role. After all, Universe was working very hard, to sculpt such beautiful masterpiece from energy and molecules, through **Intelligent Design**, miraculous in every way.

There is no such thing as Us and the Universe, **Universe is within Us**, de facto. Human performance is also producing an imprint of energy, yet, we don't know the extend of this interaction. Universe, by all means, is playing by the rules, never contradict its principles, Universe is progressing by implementation of simple rule: everything I've got is for the purpose of perpetual progression, yet, the very same paradigm shall be implemented by human society, including ethical demands, as well as projection of attitudes.

Science of perception shall implement ethical essence, yet, this process will give society unlimited potential, in terms of solutions and understanding, as well as proper implementation of knowledge.

Infinity manifest not only perpetual progression, but an ability to extend projection through accessing fragmented Universal progressions at any point. As I suggested on many occasions, one of the solutions to bypass cooling down Universe, is to go back, and access a sequence of the past, because past and the future, are virtually the same phenomenons.

In the physical world, as well as perception, which is based on time, there are mechanisms, and events, which seem irreversible. Yet, in the sequential model, word limitation is no longer applicable. Sequence is, de facto, a potent system, based on creative logic, which manifest **paradigm of maximizing potential** in every aspect, instead of posing barriers.

In a sequential paradigm, past and the future are the same, yet, defined by fragmented transitions, transformations. Universal, multiuniversal paradigm of self-preservation and performance is embedded into the human instinct, data within biological DNA. Provable by multiplications of awareness, as well as performance of the Universe, brilliantly smart, logical, Intelligent Design ... .

What is turning the wheels of a bicycle ? Intelligent Design, mind, de facto ... .

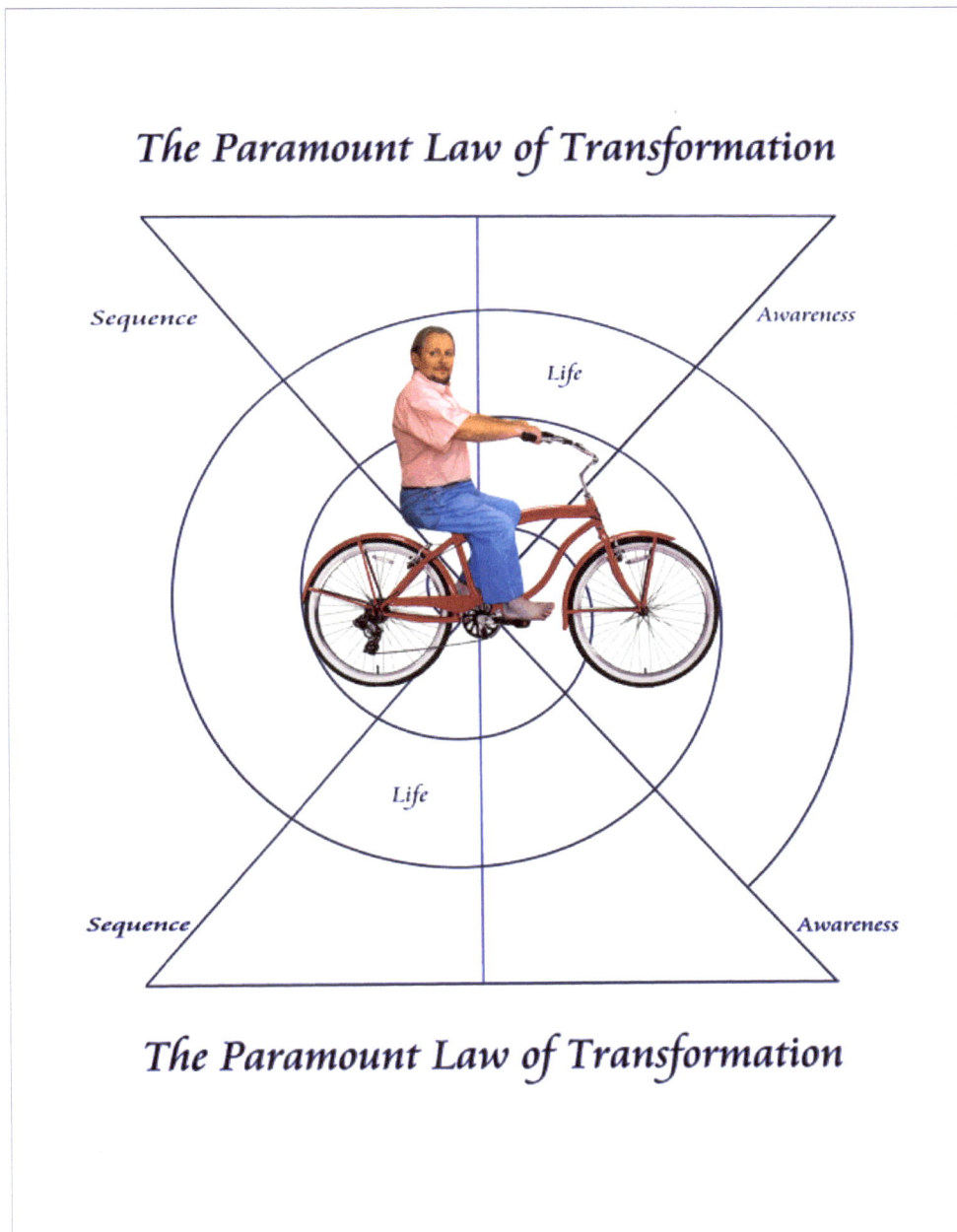

The Paramount Law of Transformation

Sequence — Awareness — Life — Life — Sequence — Awareness

The Paramount Law of Transformation

The Paramount Law of Transformation

Projections vs Universal Performance

Universe is made for performance, same as man, **The Paramount Law of Transformation, Biological Blueprint of the Universe** is proving universal quality with regard to progression of projections in every fragmentation of sequence.

What you see, with regard to the improved Camera Obscura, **George Eastman's** roll camera, you could apply to the Universal projections, yet, not made, but capturing an instant, which is unique, well, the most unique, yet, just singular element of the Universal projections, for another part, actually infinite fragmentation of the whole, which is developing its projections perpetually, through latent image, molecular, as well as energetic progressions.

Leonardo da Vinci along with other scientists, more notably Chinese, Europeans, precisely defined projections in his masterpieces, as well as other artists.

George Eastman did something very special, he invented apparatus (improved Camera Obscura) which can, virtually, access sequence at any point (some will say time) and this is the most appealing aspect with rwagrd to the art of photography, and is indicating, that sequence is potent in therms of accessibility.

Camera Obscura is projecting reflections, while George's Eastman's camera is accessing projections, with an ability to projecting images back. Actually, at some point, human will be able not only to look and hear, but interact with anyone at any fragmentation of existence, same as rolled film, exposure after exposure.

The art of projections, Universal projections, is officially open for the public.
Let's roll that babe … .

According to The Paramount Law of Transformation, Biological Blueprint of the Universe, human brain is also projecting imagery, projections of reflected light, yet, projections go well beyond what we experience through senses and awareness.

Essential tool in Divine toolbox is light, yet, spectrum of light is another matter, when it comes to molecular, as well as energetic projection. Human brain, same as the Universe, is projecting imagery externally, and perhaps both interactions define Universal projections … .

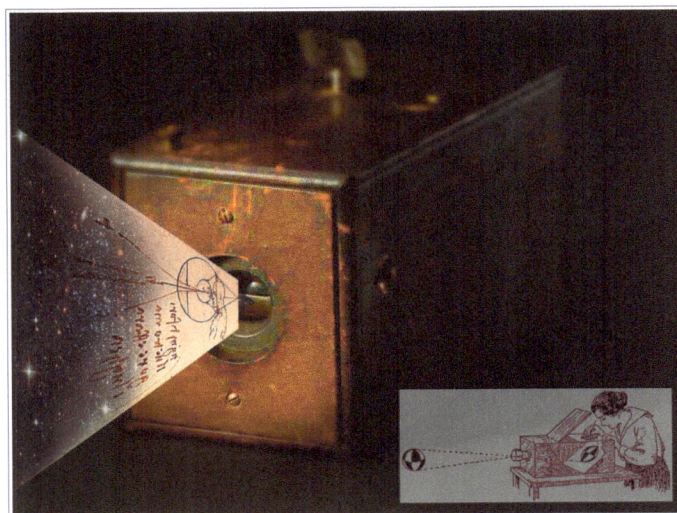

The Paramount Law of Transformation

Eloquence & Elegance of Intelligent Design
Fibonacci Sequence vs Primes vs Universal Morphogenesis vs Universal Metabolism
Gravity: Hot Light vs Cold Light

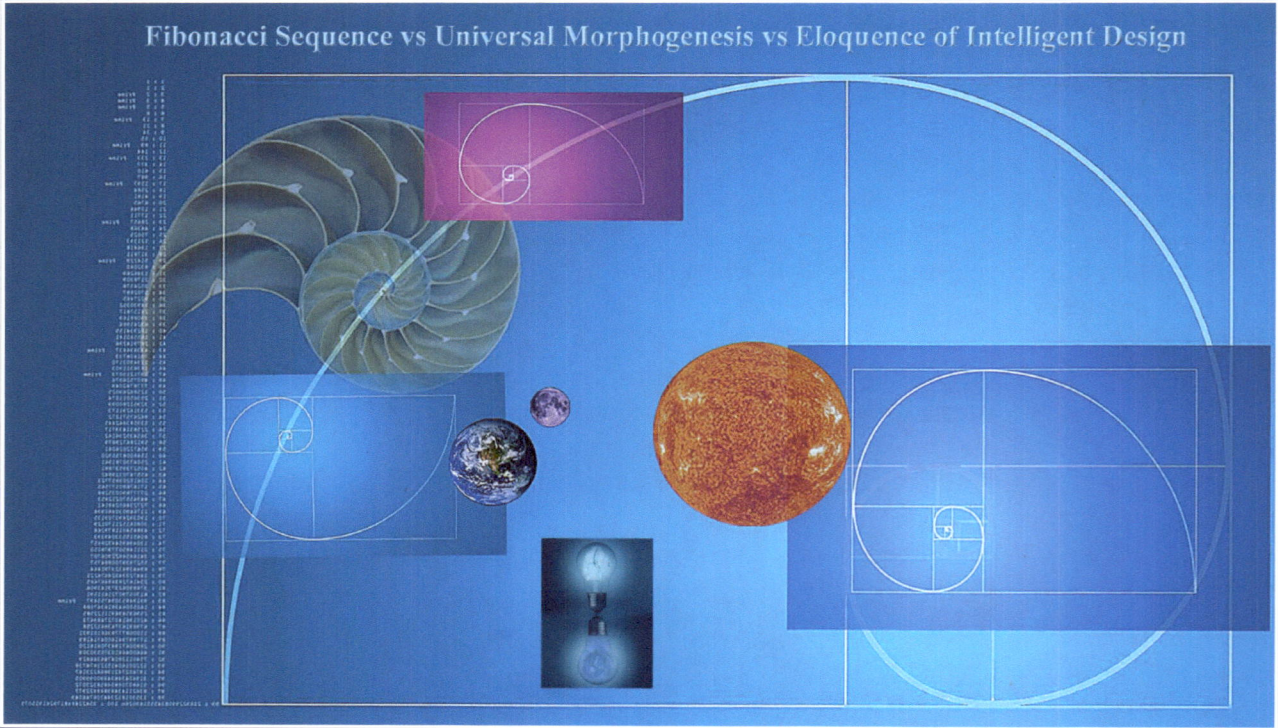

According to **The Paramount Law of Transformation**, Fibonacci sequence mark the beginning of a new sequence, based on **Universal Morphogenesis** via **Biological Blueprint of the Universe**. Spiral design is not only apparent in nature, fauna and flora, human **DNA**, but throughout the Universe. Human, according to the law of compatible opposites, mark the beginning of a new sequence as well (new life).

Motion of the Universe, local, as well as global, is reflected in spiral design: nature, Nautilus, flowers, structure of human DNA. The phenomenon of data is obvious, yet, data organized in such elegant wrapping such as 3D, and the sequential numeric system (for example Fibonacci sequence) doesn't leave any doubt about existence of **Intelligent Design.**

Data vs organized and elegantly fragmented elements, which perform in sequential order, represent truly advanced method to maximize performance, maximize potential, de facto, at the same time, minimizing friction, according to the anticipated plan, embedded sequential data, within every man, entire spectrum of elements, energies, visible, invisible to the naked eye.

Spiral design, translated through Fibonacci sequence, is reflected universally, including in the motion of the **Solar System, Milky Way** (through slow, yet, spiral motion toward the Sun), and obviously entire Universe, where everything is in motion.

In addition, Nautilus illustrate progression of projection with regard to sequential fragmentation, simple, yet, eloquent, as well as excessively elegant cellular progression, de facto, morphogenesis, where symbiosis of the design, logarithmic and geometric is truly breathtaking.

Fibonacci sequence represent progression of projection, the beginning of a new sequence, subsequently we shall ask essential question: does primes within Fibonacci prgression represent approximation of initiation of a new sequence ? If yes, than we will be able to precisely map the locations of Vortex of Matter in the Universe. If not, no harm done.

Fibonacci sequence, by the very nature, of implementation of the Intelligent Design, is triangular (as indicate illustration, as well as sequence, both numeric and geometric), yet, by studying **Biological Blueprint of the Universe**, proportions of human being, we can further calculate approximation, with regard to the **3D** shape of the Universe, as well as expansion (growth), cellular computation, approximation indicating sequential orientation of progression (at which point Universe is performing).

Universal Metabolism, according to **The Paramount Law of Universal Transformation, Biological Blueprint of the Universe** is defined, de facto, by temperature (illustration attached), within Universe as we know it, as well as in human physiology. Illustration also provide essentail indication about gravity within interaction: hot light vs cold light (light vs dark matter.

(picture source: Wikipedia)

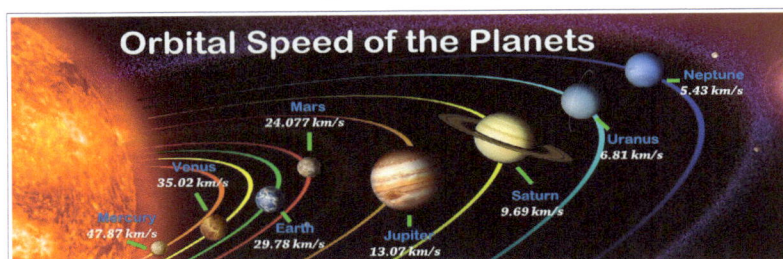

157

```
 1 : 1
 2 : 1
 3 : 2      Prime
 4 : 3      Prime
 5 : 5      Prime
 6 : 8
 7 : 13     Prime
 8 : 21
 9 : 34
10 : 55
11 : 89     Prime
12 : 144
13 : 233    Prime
14 : 377
15 : 610
16 : 987
17 : 1597      Prime
18 : 2584
19 : 4181
20 : 6765
21 : 10946
22 : 17711
23 : 28657     Prime
24 : 46368
25 : 75025
26 : 121393
27 : 196418
28 : 317811
29 : 514229    Prime
30 : 832040
31 : 1346269
32 : 2178309
33 : 3524578
34 : 5702887
35 : 9227465
36 : 14930352
37 : 24157817
38 : 39088169
39 : 63245986
40 : 102334155
41 : 165580141
42 : 267914296
43 : 433494437    Prime
44 : 701408733
45 : 1134903170
46 : 1836311903
47 : 2971215073    Prime
48 : 4807526976
49 : 7778742049
50 : 12586269025
51 : 20365011074
52 : 32951280099
53 : 53316291173
54 : 86267571272
55 : 139583862445
56 : 225851433717
57 : 365435296162
58 : 591286729879
59 : 956722026041
60 : 1548008755920
61 : 2504730781961
62 : 4052739537881
63 : 6557470319842
64 : 10610209857723
65 : 17167680177565
66 : 27777890035288
67 : 44945570212853
68 : 72723460248141
69 : 117669030460994
70 : 190392490709135
71 : 308061521170129
72 : 498454011879264
73 : 806515533049393
74 : 1304969544928657
75 : 2111485077978050
76 : 3416454622906707
77 : 5527939700884757
78 : 8944394323791464
79 : 14472334024676221
80 : 23416728348467685
81 : 37889062373143906
82 : 61305790721611591
83 : 99194853094755497    Prime
84 : 160500643816367088
85 : 259695496911122585
86 : 420196140727489673
87 : 679891637638612258
88 : 1100087778366101931
89 : 1779979416004714189
90 : 2880067194370816120
91 : 4660046610375530309
92 : 7540113804746346429
93 : 12200160415121876738
94 : 19740274219868223167
95 : 31940434634990099905
96 : 51680708854858323072
97 : 83621143489848422977
98 : 135301852344706746044
99 : 218922995834555169026n 100 : 354224848179261915075
```

The Paramount Law of Transformation

Great Exit Mode vs Procreation of the World … .

Human civilization is blending, races, nations, in more or less coordinated fashion, fair as well, yet, is world is blending through technology, transportation, accelerated rate of expansion with regard to frame of awareness, and obviously science. Undeniable truth.

According to **The Paramount Law of Transformation, Biological Blueprint of the Universe**, de facto, everything in Universe, de facto, human society is reflecting identical patterns, aware or strictly molecular, living, yet, unaware, which by the way are progressing from the very source of Universal projections.

Human civilization is blending, Universe as well, and at some point two galaxies, Milky Way and Andromeda, will blend, yet, this process is already initiated.

Once human civilization will sufficiently blend, at the same time combining all assets, ethical spine, scientific awareness, physical abilities, into one race, powerful self aware, civilization will experience Great Exit Mode, having at the same time scientific resources, which could we describe in modern terms, as miraculous.

As two galaxies will blend into super galaxy, powerful and potent toward procreation, human civilization follows the very same pattern and **Biological Blueprint of the Universe** does provide sufficient platform to make such assumptions. Yet, human can follow natural and logical progressions, or not.

Blended civilization will be able to replicate its own progressions, and projections, actually this process is already initiated. How Magnificent, indeed … .

Worlds procreate its own image, the imprint of projected sequence is similar, yet, different, unique … .

The Paramount Law of Transformation

vs

Biological Blueprint of the Universe.

Neutrino.

3I Rule. Particles can learn beyond adaptation … .

Experiment with **Neutrino** detector is pushing toward broader understanding of the Universe. Universe is progressing toward projection, which its subsequent reincarnations are showing less mass, decrease in terms of mass.

Molecular world is essentially coupled with energy, it's virtually the very same phenomenon, yet, defined by environment, conditions. Water can boiled via temperature, or pressure.

One of the phenomenons I am profoundly amazed, while observing the **Universe, Biological Blueprint of the Universe**, according to **The Paramount Law of Transformation**, progressions of projections, is **molecular ability to learn, progress, evolve**, de facto, shape a new, refined manifestations of reality.

Particle by itself can perform, progress, learn, yet, in certain complexities, there must be an energy, which stimulate progression beyond adaptation, but virtually learning and evolving.

We know that light does, de facto, sculpt the Universe, as we know it, yet, **"background energy"**, pre-programmed data is essential to perform, push toward complexities, which look like spontaneous, yet, progressions are, de facto, the result of data embedded into the particles, which are formulated to perform under certain conditions.

Massless particles, Neutrino perhaps, play that role, and it seems, Neutrinos are perfect fit to do just that, because there are virtually no boundaries or limitations, which would interfere with molecular fertilization of matter gathered within and beyond the Earth vicinity.

Human, according to The Paramount Law of Transformation, Biological Blueprint of the Universe, manifest virtually **recorded data of all Universal phenomenons**, since the Initiation sequence, yet, human is far more than just picking into the past, by recorded information within cellular memories, human is, de facto, a living, aware, performing, progressing toward aware energy, performing vision of the Intelligent Design, projection into the future, which is possible by the very nature of brain functionality via bridging past as well as the future, into sequential fragmentations, blended and manifested in seamless flow, waves, if you will, progression of projections.

We can learn, by studying respectufully, and humbly ourselves, about the past, as well as the future. Human body (functionality) represent past, while brain, mind is performing most of its tasks, within vicinity of the future, local, as well as distant. By this very function, we can visualize functionality, as well as progression of human development, which is navigating toward energy, progressively taking off layers of mass, maximizing potential of the aware energy, yet, progressing according to the paradigm of transformation data, to the very source of existence, which is molding Universe, since the Initiation sequence.

Universal spontaneous progression is defined by the paved paths, in which you could participate, while traveling from point „a" to point „b", yet, by choosing unpaved path, the results are uncertain, often, well, almost always, the result is „late for dinner", or hours behind the schedule. This approach in Universal scale could be even more profound. Let's than follow paved paths, Intelligent Design, it's smart, safe, breathtaking, wonderful … .

The Paramount Law of Transformation. Biological Blueprint of the Universe. Neutrinos.

The Principles of Intelligent Design
3I Rule (Id=Ip+Ie)
Intelligent Design = Intelligent Particles + Intelligent Energy

The Paramount Law of Transformation

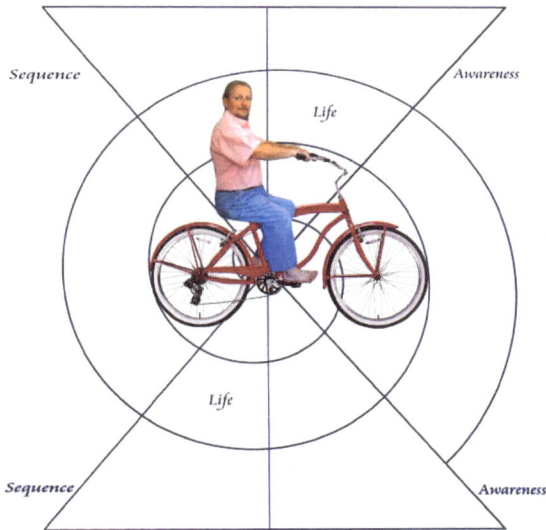

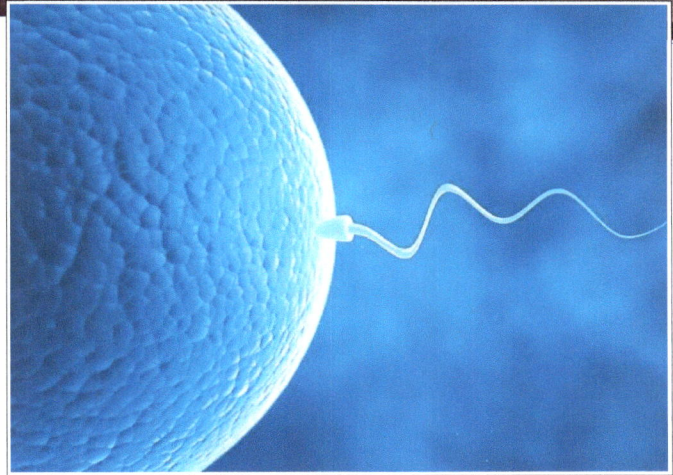

Sequence   Awareness
Life
Life
Sequence   Awareness

The Paramount Law of Transformation

(Neutrino detector, egg fertilization image: Wikipedia)

The Paramount Law of Transformation
Universal Progression of Compatible Opposites

Complete History of the Universe vs Biological Blueprint of the Universe

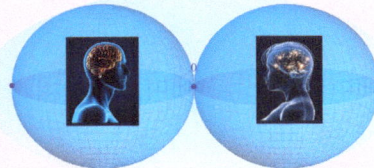

$$1\wedge + 1 = \infty$$

Energy + Inanimate Matter = Living Matter
(Hot Light + Cold Light = Initiation of Perpetual Universal Progression)

Initiation of the Universal Sequence through Compatible Opposites

Transition of Living Matter into Aware Matter and subsequently
into the Aware Energy.

Biological Blueprint of the Universe

The Paramount Law of Transformation

Biological Blueprint of the Universe
Memory: Molecular-Kinetic Memory vs Interactive Memory
Mathematics vs Aesthetic Paradigm
The 8th Cycle of the Creation

Magnificent Equations

Memory: molecular-kinetic memory vs interactive memory.

Molecular-kinetic memory vs interactive memory explains progression of Universal sequence. Within Universe all elements are precisely engineered to carry kinetic-molecular memory, data, de facto, activated via spectrums of light and temperature.

Inanimate Universe, prior to initiation sequence, cold light, background energy, dark matter, is activated via initiation elements, yet, most profoundly, energy according to **The Paramount Law of Transformation, Biological Blueprint of the Universe.**

Within Universal progression, chain reaction is activated, yet, more precisely kinetic-molecular memory, data, which is still populated by light spectrums, neutrino for example, day and night.

According to **The Paramount Law of Transformation, Biological Blueprint of the Universe,** kinetic-molecular memory progressed into interactive memory. Kinetic-molecular memory is also present within human, biological functionality, since awareness is not engaged, yet, performance is maintained. Both, kinetic-molecular memory, as well as interactive memory, are essential with regard to performance, and progression toward higher order.

The process of engineering of the Universal progression, prior initiation sequence, **The 8th Cycle of the Creation**, illustrate the preparation stage, prior to chain reaction. Universe is simple, yet, complex, and is performing according to the very same biological sequence as a man, including memory, kinetic-molecular, and subsequently interactive memory. Perhaps there are more memory fragmentations, blended into seamless substance, because Universe, every sequence is preprogrammed, as well as fertilization of matter via light spectrums, toward higher projections.

Light spectrum, ale tending the Earth day and night, where intensity varies (via waves), yet, interaction between neutrino, and inner core might be a key to understand gravity. Surely light interacts with data, kinetic-molecular memory of matter, which is progressing, as long as light spectrum and temperature are interacting via impregnating inanimate matter into molecular animation.

Universal spectacle, fireworks impregnate inanimate matter via light, including energies, which are without mass, de facto, the future of human progression of projection.

Another important aspect, are Universal stages. We know that light interacts with inanimate matter (Universal background, cold light). Temperature, universal oven is activating preprogrammed, precisely engineered kinetic-molecular memory, data, which miraculously progress into the interactive memory, yet, both dwell as an integral system of projection.

Human vs Universe, existence (stages):

• prior initiation stage, where all elements are in place yet, "fertilization" is still in the process (Universe prior initiation sequence, negative or neutral intensity of Resonant Waves)
• Universal initiation (fertilization), where molecules, as well as powerful energies are blended into the biological electromagnetic explosion
• pregnancy (immediately after the Universal initiation), development of the progression via cell multiplications and divisions
• birth, and subsequent stages, de facto formation of Universe via interaction between compatible opposites, light vs cold light, temperature
• subsequent stages of development, progression of projection.

It's important to define the stage in which Universe dwell at the very fragmentation of sequence, due to the fact, that we will be able to asses approximation of subsequent stages.

In addition, Vortex of Matter is parallel with the **Biological Process of Cell Replacement**. Both are identical in terms of maximizing performance, renewal, progression. Cellular automation, within human body, as well as recycling, reprogramming of the Universe via Vortex of Matter are fantastically similar phenomenons.

Mathematics vs Aesthetic Paradigm

Recently I've read interesting articles written by mathematician, brilliant, sophisticated, yet, at some point he is saying: "mathematics doesn't have to be elegant". As I wrote in my articles, math and geometry represent interactive narration, yet, translated in to two different languages. If math can not be translated into the geometrical eloquence, than something is missing, even if it works for the moment.

Dear friends, entire Universe represent progression of projection, 3D, Alfa-Dimention, from which progression of possibilities is developing from less complex to the very complex formulations. Aesthetic paradigm in Universe is profoundly embedded, yet, man, is encapsulated within, as well as externally with aesthetic perception, manifestations, aware as well as unaware.

Matter is populated through spectrum of light, and it seems, the process, data embedded within, penetrate visible and invisible world, **Molecular Communication and Neutrino** (chapters from my book **The Paramount Law of Transformation. Biological Blueprint of the Universe.** The Law of Everything), are stating that aesthetic paradigm is essential, built-in in every process of transformation.

Elegant projection is not an accessory, but fundamental quality, with regard to blended fragmentation of the Universal progression. Why ? Because elegance is seamless, represent physical quality, with regard to practical solutions, besides perceptual fiesta. Aesthetic paradigm, elegance is, de facto, at the foundation of **Intelligent Design.**

The Paramount Law of Transformation
Mathematics vs 3D vs Molecules vs Energies vs Reality

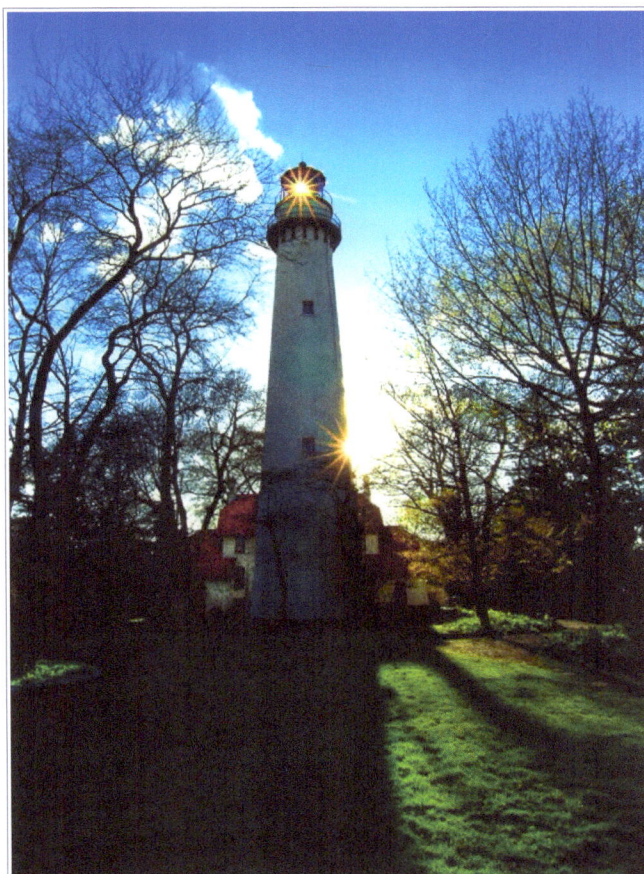

Mathematics, algebraic language is flat, any number represent simplified translation from 3D (mind vs 3D projection vs idea).

Universe is Geometrical, and then translated into numbers and equations. This is the process of rediscovering reality, as well as perceptual expansion of awareness (I wrote about it for more than a decade).

Number is flat, even equation represent flat, non practical property, unless will be translated into 3D projection, yet, the very process is reversed, 3D projection, any projection, is Universal, and then can be translated into numbers, not vice versa, numbers which dwell within 3D space.

As I have indicated in my articles, dimension is irreversibly interconnected with 3D.

Number is, de facto, flat in 3D space, yet, projected via intelligent process of rationalization from 3D to numerical sequence. This is the classic example how **Intelligent Design** simplify process of the design, across the Universe, within human mind (mind performs, propagate projection within 3D through electromagnetic waves).

I would like emphasize, once again, 3D Geometric progression of projection is integrated, yet, numbers and equations always dwell within **3D, Alpha Dimension,** the source and platform in which mathematical language can perform.

Number is flat, yet, gracefully performing within 3D Universe, Intelligent Design.

Reality. Reality is only one, singular, undivided, because represent progression from Universal sequence, since initiation, and then optimized into perceptual, molecular, unaware and molecular aware. Reality is expanding along with 3D awareness, the very frame of perception, we define as knowledge, sophistication. Yet, Reality reveal its secrets through progression of awareness and subsequently projections, yet, projections oscillate within energies.

If there were more realities, molecular, than variety of realities, not connected, would represent chaos (human invention), inconsistent Universe, which would not perform, as it does according to **The Paramount Law of Transformation, Biological Blueprint of the Universe.** Singular Reality is progressing into variety of projections, yet, represent blended progression from the very source, fragmentations, projections … .

Molecules. Molecules are indestructible, yet, transformable into other physical imprint of data. If particle disintegrate, than, de facto, data transformation is at play. Molecules, any molecule represent energy. That's all it is, our entire integrated world is oscillating within energies, from simple to complex progressions and subsequently projections.

Molecules vs energy. As I have indicated in previous articles, every molecular data is, de facto, progressing toward energy, potential or real. Molecules, physical world is formulated toward performance, precisely engineered molecules, able to change its state into energy.

I Believe in Intelligent Design and I know that you and I we are the projection of the most wonderful performance, Theater, Divine in every detail and manifestation. I am grateful for the opportunity … .

The Paramount Law of Transformation
Human vs Positive Compounding Energy vs Negative Compounding Energy

The Paramount Law of Transformation
Biological Blueprint of the Universe vs Energy
The Law of Everything

Human is the most **sophisticated progression of projection** within Universe, and via reaching the stage of aware energy, human is virtually interacting with the Universe, yet, human has the ability to shape Universal reality by multiplication of aware energy. In this very instance, the performance of Universe is also determined by human performance (classic law: action vs reaction).

We shall not disregard this notion, for the sake of the future, as well as the process of designing positive compounding vs negative compounding energies, profoundly important, and defining the future of human race.

I would like to emphasize that human performance, in terms of positive energy vs negative energy (ethical standards) is resonating via energies, most sophisticated, and this process is able to transform the Universe. How extensive this interaction is, well, it is hard to determine, yet, interaction does exist, resonate.

Virtually everything within Universe, as we know it, is energy, data, de facto. Density play important, if not defining role, yet, we live an thrive in varieties of energies, precisely distributed and managed. Human represent Universal projection, yet, transforming still, into aware energy, which logical subsequent Universal development, yet, it is, de facto, a journey to the very source of Universal progression.

Human manifest subsequent progression of kinetic energy into kinetic interactive energy, energy, which if progressing at this very moment.

The Paramount Law of Transformation
Vortex of Matter vs Resonant Waves vs Initiation Sequence

Vortex of Matter is creating **Resonant Waves.** It is possible that by releasing energy and data into inanimate space, subsequent Initiation sequence is taking place. The intensity of Resonant Waves, at certain point of sequence, activates chain reaction, similar to the **Initiation Sequence** of our own Universal, profoundly potent World.

Described process is compatible with **The Paramount Law of Transformation, Biological Blueprint of the Universe, The Law of Everything.** Resonant waves are present within Universe, as we know it, yet, within human functionality, brain electromagnetic waves, heartbeat, human electromagnetic field.

Resonant waves dwell within Universe, as we know it, de facto, manifested within intellectual and physical functionality of every human being. The Paramount Law of Transformation, Biological Blueprint of the Universe represent the essence of **Intelligent Design**, yet, Universe, in all occurring phenomenons is mirroring essential, fundamental laws, paradigms of progressed projections.

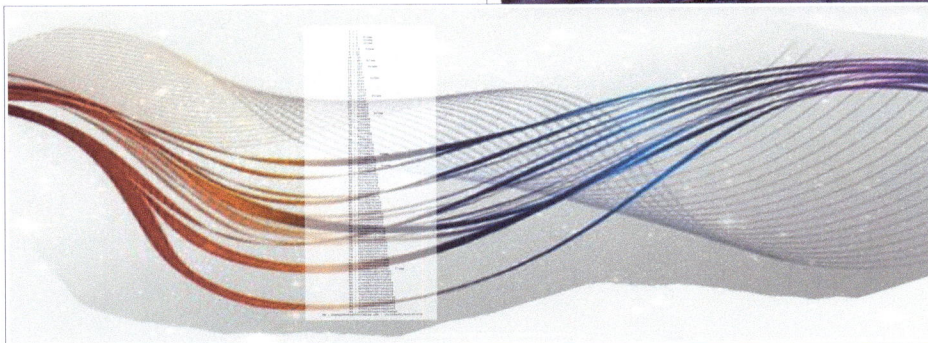

The Paramount Law of Transformation
Biological Blueprint of the Universe

Space Program sinc3 1452 …

My fascination with the man, carried me to the Universe …
Quantum Data

Great Unification encapsulated. Quantified Symmetry…

Space Program sinc3 1452. Hop in for a ride, and watch the Universe projecting its beauty … .

While Quantum science is progressing, and producing interesting applications and solutions, is still dealing with minimum data available.

Human, de facto, represent Quantum science, complete and effortlessly implemented throughout progression of projection, entire data of Quantum physics, Quantum Mechanics, Quantum Computation, Quantum Physics, all is here, Quantified DNA of Intelligent Design, Divine textbook within, and the working, performing perfectly model of an idea.

Dear friends, minimum data is not enough, its like analyzing the Universe, while looking only at male or female. Even if it works, at some point, you will discover that it won't feet in big picture, or you will create something unlike the world we know … .

Quantified Reality vs Improvisation
Space Program sinc3 1452

Improvisation, within permissible frame of performance, represent singular example of chaos, quantified reality bis, de facto, free to perform, yet, its progression is precisely defined … .

Quantified Sequential Waves

Universal Wave Paradigm is compatible with quantified progression, human heart, brain waves, de facto, reflecting Universal pulse, waves of light. In addition biorhythms are also mirroring sequential waves.

Human kinetics, biological mechanical system is compatible with wave paradigm, while youtake step forward and backward …

Frequency and the Initiation Sequence

Photoelectric effect is present in Biological Blueprint of the Universe, where at certain frequency of light human is active and while frequency changes, human is at rest. Ideal example of resurrected Space from a state of hibernation, where at certain conditions, threshold frequency, world is called to life … .

Quantum Percpetion

Quantum perception makes sense only if entire picture is projected.

Quantum Science vs Gravity
Space Program sinc3 1452 ... .

DNA, quantified sequence, with 3 billions fragmentations, design is progressing according to the principles of the design. Without big picture, Quantum science, calculations would produce strange results.

DNA shall be protected by Divine Law of the Universe, and forbid experimentation of any kind. Gravity represent essential engine of molecular unification through motion, gravity is localized and global ... .

Yet, everything within the Universe is energy, potential and kinetic ...

Quantum Library vs Geometrical Projections

Quantum exploration makes sense only if quantum library can be translated into geometrical projection ...

Quantum (the essence):

* singularity dwells in whole, yet, all dwell in singularity = I am in you, you are in me ... .

Quantum Science vs Scrolls of Miracles ? Yes ...

Quantified progression of projection, if properly formulated, as it is, represent Scrolls of Miracles within Universal Library, where singular quantum is in all, and all is within singular (here we have another example of science, which is formulated philosophically, precise science, and beauty of vision ... .

Beauty of simplicity in quantified unity.

I wrote about it in USA (2006), singularity dwells in whole ,yet, all dwell in singularity = I am in you, you are in me.

There is nothing that can surpass this beauty, nothing better will be ever invented, pure essence of unification and progression of projection we can enjoy without limits.

World is abundant with beauty, breathtaking science, which has heart, gentle and sensitive, to share all of that so life can progress from paradigms of science via paradigms of ethical, philosophical, artistic, sensual phenomenons.

If we will be able to translate this paradigm to social awareness, than world will be transformed in an instant, a true path to happiness, knowing that social quant5ified reality eloquently suggests, that each quant is a key to another ... .

Initiation Sequence and Vortex Localized and Global … .

Program sinc3 1452. Hop in for a ride, and watch the Universe projecting its beauty … .

Initiation sequence occurred in one location, via interaction between hot and cold light (light vs dark matter: compatible opposites). From global vortex, due to decrease of initial energy progressively emerged localized vortex.

There is a limit or boundaries of the Universe, as we know it, where vortex of matter is limiting expansion of the Universe. Without spiral motion on the Universe, life would not emerge, at least in a form as known as reality, perception, Earth, Universe.

Vortex of matter could be sufficiently studied by looking into the human eye, vision system, projection properties of human body, transformative beauty of light.

Human brain is emitting super energy, compatible, yet, aware and fundamental, with regard to building blocks of Universal progression of projection. Human brain is producing energy which defy molecular phenomenons and properties. Science for Peace … .

Is space infinite?

Is space infinite, or just really big? Even if it has a fixed size, the universe might have an infinite number of locations unless, that is, if you take into account …

It depends how you measure it, with big wheel or small.
Infinity represent Universal transformative power, transition from one state to another without compromising progression … .

The Paramount Law of Transformation. Biological Blueprint of the Universe.
The Law of Everything: Great Unification encapsulated. 3D … .

In Universe, as we know it, there is only 3D projection, and while talking about 1D, 2D, we are talking about perception not science.

Every molecular projection require demand space, which is by the definition 3D, yet, other dimentions are nothing more, than progression of projection from 3D, unless aware energy can penetrate 3D, and all subsequent projected dimensions.

3D: if you have any doubts, peacefully accessorize your vision with appropriate magnification, what will you see ? 3D, stubbornly starring at, well, you, from all angles … .

Symmetry

Symmetry represent progression of transformative power of Intelligent Design. Without Symmetry, progression of projection would not be possible it also suggests that multi-sequence is projecting in all directions, and this revelation also suggest multidimensional entities (Universes).

Symmetry indicate, that Universe, as well as all parallel Universes create, de facto, 3D interwoven plaits (tress).

I already explained this notion in previous articles, yet, Symmetry require more attention. We know by perceptual, as well as empirical, and intuitive power of our awareness, that Symmetry is the result of motion / transformation (Vortex created circle as well as derivatives).

Within Human is embedded, equally, Symmetry, as well as so called "broken symmetry", which, de facto, is progressing into another, subsequent symmetrical progression of projection.

Asymmetrical Universe is possible, yet, rendering symbiosis between 3D geometrical projection, as and numeric/algebraic, would be inefficient, and produce unnecessary, enormous complexities.

Symmetry illustrate the power of simplification, as well as infinite creative power… . We can describe Symmetry as an essential transformative progression, a path embedded within Intelligent Design.

Symmetry, essential beauty of Universal progression of projection: Universe vs Human, both share identical characteristics seamlessly designed and performing. Brilliant eloquence.

What I cherish about Symmetry: The Beauty of Philosophical and Empirical Science, where entire Intelligent Design is encapsulated … .

Attached image may contradict the notion of Symmetry, yet, the phenomenon is profoundly incorporated into the design and transformation (I call it a hidden symmetry), which is propelling progression from within … .

Improvisation

The Paramount Law of Transformation. Biological Blueprint of the Universe.
Organic Character of the Universe: Parallels.

Universe is a living organism, very much organic, through molecular and energetic progression of projections.

Within Universe is encapsulated entire science, yet, the master equation is, well, Man, who define all, with only few, rare, exceptions.

Universe is breathing through its molecular lungs same as Human, where space is filled with energy (oxygen), which interact and allows progression of projection to perform.

Space, prior to Initiation sequence was filled with inanimate matter, which still dwell and perform in front of our eyes, along with spectrum of light … .

Music vs Fine Art vs Universe vs Great Unification
Great Unification encapsulated … .

Ravell's Bolero eloquently illustrate the science of Paramount Law of Transformation. Music, in this instance, simple, yet, beautiful composition manifest Resonant waves, compatible with human resonant waves. If you would play Bolero forward and backwards (I've mention about in the past), than resonant waves of musical composition are reflecting how Universe is formed.

Tam, tam, tam, tam, tam, resonant waves are intensifying, frequencies, tam, tam, tam, tam, tam, culmination ! That's the defining moment, when inanimate Universe is impregnated with energy and matter (first word). Exact the same kinetic mechanism is observable in human interactions, where energy bis translated into kinetic energy, and both play great composition of life.

Yet, where Bolero ends, Universe begins to live and that's the Great Unification is taking shape. While combining Ravel's Bolero with Escher's masterpieces, geometry and algebra, nearly entire picture emerges, simplicity in complexity.

Biorhythmic Resonant(Biological) Waves Principle

What is the nature of time, well, as I said previously, time represent an elegant accessory, yet, more precise description about Universal processes, which take place, is Sequence and apparent;y fragmentations. Nature of time, measuring a motion of celestial spheres and cycles.

At some point, man will perform and pay more attention to the inner paradigm of sequential transition (inner clock) so called biorhythm, which represent Biorhythmic Waves, Biological Resonant Waves (Biological Wave Principle) … .

Universe is speaking with gentle voice: **I dwell in You, You dwell in me … .**

Human came from stars and will return to stars. Human society is transforming vital ingredients of existence into wasteland: nature can dwell without man, man can not dwell without nature.

Instead of consumption, uncontrollable and unsophisticated, human civilization shall put all available resources, science, talents, ideas, spiritual centers at work, immediately, toward self-preservation.

We, humans ought learn the joy of life, by sensual, as well as intuitive, spiritual engagement with miraculous, natural forces, which were working exhaustively to achieve the art of existence.

How many men on Earth appreciate the gift of aware existence, how many. The purest joy we can experience is through realization of being, knowing, performing in an interactive network of identical heartbeats, translated eloquence of Intelligent Design in your own DNA.

Space dwell within you, and is telling us, do things differently. Live modest, yet, fulfilling life and learn, how to speak with energy of life, with every breath you take, the very energy which carried life on its wings in vast space, and the very same energy of life which will give man wings to fly once again, yet, not flee from place, which was meant to be a source of our pride, our home, Earth… .

How many people enjoy sunrises, existence of light, illumination, visible and invisible, which is tending the Earth, nature, human bodies, day and night. How many people can feel energy of beauty, embedded into every living molecule, even if it can't speak, yet,it does telling great Universal story via language of tr5ansformative power.

There is a joy of having, yet, the joy of being is so profoundly beautiful, and wasted, in every missed opportunity to be better, more meaningful to experience joy and opportunity, and at the same time, making every effort to avoid disruption in the network of hearts.

Universe is speaking: **I dwell within You, You dwell within me … .**

Do not transform the world into the wasteland, but abundant garden, where flowers and spirits are blooming … .

Everything is made for human enjoyment, fulfillment: colors, shapes, stars at night, senses, love, compassion, the richness of the world is for you, yet, artificial incentives override the essence and purpose.

Live abundantly, yet, start your day by appreciating the gift, gift of life in interactive network of hearts and thoughts, and be perfect by the willing act toward self-perfection.
Universe is speaking with gentle voice: I dwell within You, You dwell in me … .

Symmetry … .

An illustration project progression of Symmetry, Supersymmetry and progressivelly „breaking" the Supersymmetry. 3D projection illustrate transformation of forms, with regard to molecules and energy as well. The system is complete: 3D projection, Symmetry, Supersymmetry, „breaking" Supersymmetry via transition into progression of Symmetry, Intelligent Design according to The Paramount Law of Transformation. Besides completness, the system of transformation reflect another quality with regard to Intelligent Design, the Permanence of Aesthetic Projection within … .

Supersymmetry vs Biological Blueprint of the Universe

While studying molecular automation, cellular computation in USA, I realized that the Symmetry progressed into the Supersymmetry, and subsequently „breaking" Supersymmetry. Terminology „breaking" Supersymmetry is slightly misleading, due to the fact, that Supersymmetry is progressing into the new projection, which has its sequential source in Supersymmetry.

Submolecular level of progression is no different, unless, human technology will produce a new, unknown particle(s) or fragmentation of Universal particles. In this instance, human race can reach the situation of disrupting molecular integrity by parallel forces, as Vortex of Matter (Black Hole).

Once this happens, the Solar System, Milky Way Galaxy, biological forms of life, would stretch like a glue or jelly, and than, the entire structure of Intelligent Design, would be invited for the ride, (possibly reset).

Supersymmetry is profoundly projected in Biological Blueprint of the Universe. For example, brain is symmetrical, yet, fragmented by the design in terms of functionality (right vs left hemisphere), compatibility of opposites with regard to sexes, progression of sequence, since initiation, biological transformations.

Geometry is essential in Universal progression, yet, unique illustration of the process is encapsulated in M.C. Eshers's genius. In this very location is embedded the power of the Universal progression of projections, symmetry vs "breaking symmetry", which is still progressing into transformative beauty as well as practical (the state of mind of Leonardo da Vinci).

In M.C. Escher's masterpieces is encapsulated the essence of symbiosis between geometry and algebra, seamless and, it seems, effortless transitions, without creative boundaries or limitations, yet infinitely creative and practical Intelligent Design.

I have not seen, anywhere, the purity of Universal progression of projection, as is manifested in M.C. Escher's works, yet, I can feel it, touch it in my mind, senses as well as via the power of intuition, the very breeze of the Universe, which dwell in my soul … .

Neutrino

Just bring on board, good human nature, positive attitude, and you are set … .

Neutrino is not a particle, yet, energy, as I have indicated previously. Everything in Universe, as we know it, is energy, potential or kinetic, yet, Universe, molecular is populated by data, energy, very much kinetic in nature, which translate as The Paramount Law of Transformation.

Everything that is energy is emitting data, yet, Biological Blueprint of the Universe is progressing into projection of its own projection, imprint of data through aware energy, which interacts with the entire system.

Kinetic memory, interactive memory progression is parallel with progression of molecular state and energy … .

Gradation of mass define, de facto, the very boundaries between molecules and energy, via kinetic stages, via interaction of Compatible Opposites. Yet, to give entire truth a chance to speak, we could say, **Biological Blueprint of the Universe** manifest the very same progressive principle via biological stages of development. And this is the beauty of Intelligent Design …

The Paramount Law of Transformation. Biological Blueprint of the Universe. Space Program sinc3 1452 … .

Quantum (the essence of the Universe):
* singularity dwells in whole, yet, all dwell in singularity = I am in you, you are in me … .

The Paramount Law of Transformation (ῶ) Magnificent Quantum Awareness

Quantum Theory is a real phenomenon, an essential law, based of fragmentation of sequence, particles as well as energies.

Wave paradigm manifest quantum properties as well. In addition, quantum physics from subatomic phenomenons, energies, de facto, to UNiversal, represent the very same quality.

To measure Quantum, is like measuring a thought, yet, it is possible to measure some principal physical properties (length of wave). Quantum r3present scientific philosophy with regard to formulating Universal sequence, where the fundamental paradigm is the vision, an idea of the whole (without it strange creatures from the ancient Greek mythology come to mind - Gorgon or Minotaur).

Wave Law on UNiversal scale is, de facto, the largest Quantum phenomenon within Sequence in Universe as we know it. In addition, properly formulated quantum fragmentations, resonate and Initiate a new sequence, beyond Universe as we know it, a new Universal sequence beyond our own.

Miraculous initiation through resonant waves of subsequent Universal entities, is a classic and eloquent sinusoidal existential quantified wave, which vary in length, as well as intensity. Our existence is essentially defined and formulated, approximately, by sequence of wave from the Sun to Earth, I call it **The 8th Cycle of the Creation** (energy of light is traveling approx. 8 minutes from the Sun to Earth - simplified formula).

Universe, as we know it, is going through Universal Quantum sequence within a **Wave Paradigm** (**r3sonant** length represent quantified property within another **quantifi3d** property, where all are part of quantified whole, and whole is quantified part, which translate as follows: we live in a subsequent Universal sequence, yet, space, which we call it Universe, either was or will be, or was and will be sequential (this approximation is perhaps the most accurate), according to the **Quantified Universes**, which are subsequent progression of projections; same system, mechanism is implemented from subatomic/energetic to Universal and Multiverse progressions of projections.

The power of **Intelligent Design** is manifesting its genius in simplicity.

The Paramount Law of Transformation.
Biological Blueprint of the Universe.
Space Program sinc3 1452 … .

Intelligent, transformative function within Intelligent Design is shaping reality.
Biological accelerator.

All phenomenons within Universe as we know it, every particle represent an energy, either potential or kinetic, the very manifestation of **Intelligent Design**, based on transformative **DNA.**

Biological existence, well, every existence, molecular unaware, as well as molecular-energetic aware, represent progression of projection, where light manifest essential ingredient, along with temperature and space. Universe is a special kind of oven, which is based on wave pattern as well (hot, cooling off period, cold, hot, colling off period, cold).

Every singular functionality of unaware, as well as aware existence, depend on light, and is manifested via interaction between cold and hot light (compatible opposites).

Human manifest, de facto, **electro-magnetic engine** within space. Human body is absorbing energy in various forms, yet, light, it seems, is tending the Earth all the time, along with a man. All spectrum of light, visible and hidden, are interacting with human body where light is virtually carving human DNA with emb3dded transformative data.

We know, that all objects within universe, are representing potential and kinetic energy. Molecules absorb energy (molecular or higher order), which are being transformed as well.

My fascination with the subject is oscillating around transitional energetic transformation within human body, brain in particular, and a mechanism which is emitting energy via brain to the near, as well as remote space. **Human brain** is absorbing energy and subsequently is emitting transformed energy, yet, **human body, brain represent quantified accelerator, particle collider**, the facto, an essential ingredient of Intelligent Design.

My own observations indicate, that human energy, energy of thought is refracting from surfaces, yet, intelligently driven by awareness or subconscious mechanism. Human energy emitted by brain/thought is either **projected as a wave, or a beam of energy or both**, and is very much based on **Wave pattern.**

Human brain is emitting either concentrated beam of energy or wide beam/wave of energy via inner mechanism driven by awareness, as well as unaware sophisticated mechanism, which is able to interact with an internal, as well as external world.

Wave Principle are replicated throughout the Universe, within biological existence as well, due to the **Paradigm of Symmetry**; **Wave Principle** is "breaking" the Symmetry, yet, it is a Symmetry still, but the pattern vary from one transition to another. In short, Symmetry is unbreakable in Intelligent Design, yet, is progressing from a certain pattern to subsequent pattern, in order to maximize molecular and energetic potential in quantified unification, within progression of projection. Nothing is left for a chance, yet, acceptable variations, **Universal improvisation** are common.

Human brain works similarly as an energy collider, while absorbing energy, and emitting energy as well, de facto, the very basic function of Universal molecular interactions, yet, highly specialized and coupled with awareness.

Intelligent, transformative function within Intelligent Design is shaping reality. Attenuation of chaos within Intelligent Design, is achieved and associated with Symmetry as well as breaking symmetry – Wave patterns, in an organized and highly sophisticated manner.

The Paramount Law of Transformation. Biological Blueprint of the Universe. Space

Program sinc3 1452 ... .

**Universe; individuality vs parallels vs quantified essence.
Molecules; formation of physical data.**

Human, in terms of the design, manifest breathtaking solution, progression of projection. The question is as follows: human civilization, in all fantastic abundance and similarities, represent singularities, yet, different, but at the same time, not so remote in all functionalities, physical, mental, spiritual, as well as intellectual eloquence.

There are vast variations, which shape individuality, yet, one of the most profound is location of biological system vs projected light in space. This is, in my opinion, very important clue, when it comes to singularity of individuality, defined profoundly through quantified progression of projection with regard to spectrum of light, subtle molecular, as well as energetic radiation, data, which profoundly shape human existence, so similar, yet, different (locally, and even more diversified globally).

While living in united quantified Universal progression, variations, in terms of exposure of data, spectrum of light, energies, potential and kinetic, projected by the Universe, fragmented, yet, compatible data, is fertilizing human, well, all biological existence (fauna and flora and even living unaware molecular world), even stars and galaxies, with an individual imprint. The only common manifestation within Universal sequential progression, is diversity, yet, progressing within permissible vicinity of practical and aesthetic eloquence.

Yet, equential parallels of individuality, are driven by **Intelligent D3sign**, fueled by vast compatible diversities.

Beautiful d3sign of Universal progression, indeed, where individuality dwell, while propelling entire sequence toward infinity, well, I shall say, it's the essential ingredient of infinity … .

Diversity in unity toward infinite progression of projection in aesthetic, as well as practical sequential parallels. It does make sense, simple, where complexity is being reduced by seamless beauty of Intelligent Design.

All of the above is provable and observable within Biological existence, as well as throughout the Universe as we know it, and I'm sure beyond … .

The Paramount Law of Transformation.
Biological Blueprint of the Universe.
Space Program sinc3 1452 … .
Molecules; formation of physical data.

Molecules within Universe, are being formed through nuclear fusion and motion, spin, de facto, centrifuge and temperature, where fragmentation is progressing, while forming molecules from energies, including subatomic to heavy particles, yet, all molecules come from an energy and Universal-global, as well local, within vicinity centrifugal system.

Universe is turning globally and locally, as well as contracting and expanding. Identical properties are observable within human, and nearly all biological and molecular manifestations of existence.

Temperature and motion, Universal-global, and local, separate and form molecules, yet, while global Universal temperature is decreasing, motion of Universe is also slowing down, centrifuge, which populate Universe with molecules, as well as energies, are also affected by this process.

Human brain represent subsequent model of centrifuge, while forming, through quantified progression and projection, aware energy within Universal sequence.

The Paramount Law of Transformation.
Biological Blueprint of the Universe.

Space Program sinc3 1452... .

Universe; Physical data vs Rising of Awareness.

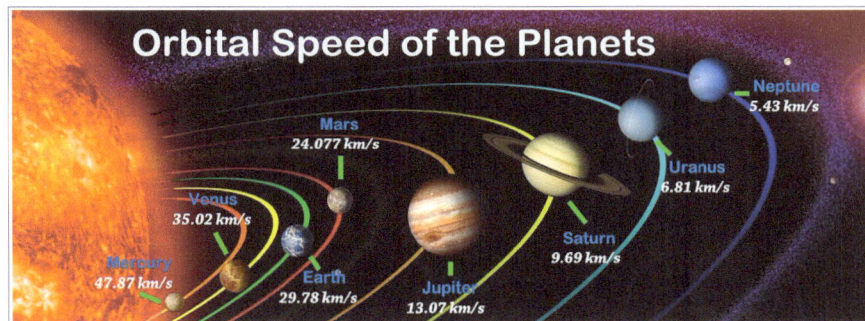

Awareness is as old as a Universal progression, since miraculous initiation sequence took place. De facto the history of awareness is the history of light interacting with an inanimate matter. While contemplating the origin of awareness, we shall analyze few essential universal qualities with regard to Intelligent Design; light, space, motion.

Light manifest the energy of transformation anchored with motion, which performs in space. The history of awareness starts in this very moment, somehow destined to reach the point where kinetic, interactive awareness arise from a static molecular awareness, yet, projection of awareness is progressing still.

At this point we can define three major types of awareness:
•static molecular awareness, where particles react with elements in organized and predictable scheme, such as spectrum of light, temperature and motion
•dynamic interactive awareness, manifesting projection from a static molecular awareness (human)
•aware energy, a subsequent progression and the facto, returning to the very beginning of the initiation sequence, the source of Intelligent Design.

Subsequently we search for a defining moment, a tipping point, where awareness is taking its shape and eloquence of understanding, beyond necessity of molecular awareness, as well as the essence of calculative performance within awareness. This is, de facto, defining moment; orientation of a biological object vs light vs motion.

In this instance, miraculous performance of awareness becomes a necessity with regard to progression of projection. Human Biological Blueprint of the Universe illustrate, eloquently, the phenomenon of awareness even in human anatomical design – brain, which is the most exposed, optimized quantified feature, with regard to spectrum of light, as well as spherical motions (global as well as within vicinity).

Awareness, by the design, was miraculously developed by a necessity, orientation, with regard to spectrum of light, spherical motion in terms of global, Universal motion, tr5nsitions, as well as local. Perhaps the very first feature of the awareness was an internal gyroscope, highly sophisticated mechanism within biological functionality.

Molecules can learn, and are formulated in such a way, somehow destined for performance. Awareness, it seems, has a tendency to replicate a distant pattern, definition, from which entire sequence was initiated, the essence and the source of transformation.

Once again, we can repeat previous statement with regard to quantified Universal sequence, and apply to the sequential rise of the awareness from a static through dynamic, interactive to aware energy of higher order :

Quantum (the essence):
•   singularity dwells in whole, yet, all dwell in singularity = I am in you, you are in me … .

Breathtaking simplicity, in every instance, Divine … .

Replication sequence, in this instance rising of awareness is also adequate, and parallel in biological, human interactions; DNA, social.

The Paramount Law of Transformation.
Biological Blueprint of the Universe.
Space Program sinc3 1452.
Rising of Awareness … .

Light along with visible and non-visible spectrum represent profound quality, perhaps the very essence in terms of existence, unaware molecular, aware, as well as aware energy.

Yet, light via pre-engineered molecules provide sequential orientation, alignment, which, de facto changes, transforms everything on its path.

Intelligent Design is based on a few very basic principles, yet, molecular world, energies, oscillate, miraculously, around spectrum of light. This very physical quality define existence, where acumen particles (molecules can learn and apparently evolve), are changing physical status quo.

While everything within Universe is in constant motion, orientation/alignment vs motion and the source of data (light) is, de facto, the paramount quality and essential transformative tool, from which Divine is managing physical world.

In addition, orientation notion, alignment vs light spectrum is also profoundly important in social interactions, spiritual as well, besides existential.

All laws within Universe, as we know it, are Universal and embedded in molecular, as well as biological sequence, yet, present within aware energy.

Universal Reference Point = Alignment of Physical objects vs the Source of Transformation; spectrum of light and associated sequential accessories such as cold light, temperature, space.

Due to the Universal symmetry, the notion of the center of the universe irrelevant, yet, biological projection, is where progression reached important, miraculous chapter, since initiation sequence took place.

In addition, spontaneous initiation of the Universal sequence was a result of resonant waves, which pounded inanimate space until initiation was possible. With regard to previous article, published last night, I would like add that molecular world and Universe there are three distinct types of **Universal Reference Points:**

* static alignment (molecular world)
* dynamic, interactive alignment (biological and humans)
* aware energy alignment (progression of intelligent, interactive awareness and intelligence.

Motion vs light spectrum leads to awareness and logic. It is an essential relationship between energy, fundamental carrier of data vs space vs motion. This is the prove, that particles learn and progress, even without interactive awareness.

Intelligent Design is intelligent throughout and it is a wonderful discovery that a chance is being replaced by the superior design is undeniable.

The Paramount Law of Transformation.
Biological Blueprint of the Universe.
Space Program sinc3 1452.
„Zero" ? Yes … .

Zero is ev3rything, de facto, represent entire sequence, the very act of initiation of transformation. Zero is defined by the aware act of progression, which has its source in an idea, progression which reached the stage of projection, unified, logical quantified sequence, perpetual, indeed ... .

The existence of z3ro, is, de facto, definiton of existence and unlimited pot3ntial... .

Perhaps zero is the defining moment in quantified Universe, as we know it.

Zero connects Universal progr3ssion of proj3ctions, beyond our own perc3ptual boundaries ...

In quantified sequential progression, where only past and the future represent 3mpirical manifestations, zero is perhaps the closest approximation to present, from which emerges quantified, unifi3d beauty, awareness, yet, zero flows, like every other molecular and energetic transformation in quantified progression of projection.

Shall w3 say, progression of projection, or perhaps projection of progression.
Both are correct, yet, **proj3ction of progression is the essence of Int3lligent D3sign ... .**

The Paramount Law of Transformation.
Biological Blueprint of the Universe.
Space Program sinc3 1452.

Micro-Cosmos vs Macro-Cosmos
Sphere vs Platonic Solids vs Power of Quantified Transformation

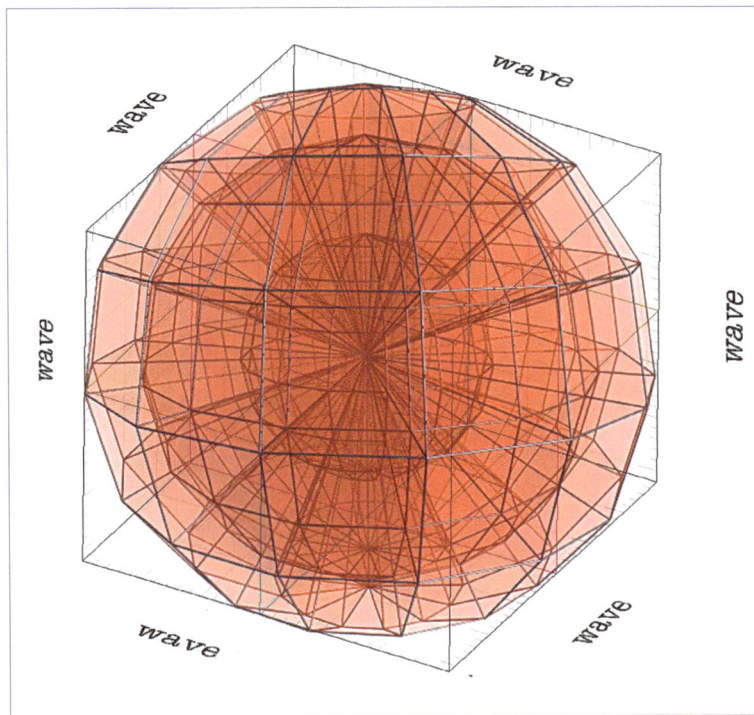

Platonic Solids are missing one ingredient, a Sphere, generated by motion between compatible opposites (through spiral motion). Sphere is borne and subsequent geometric incarnations.

In addition, Platonic Solids along with a Sphere: **6 in all**, represent simplified Multiverse, transitions from previous into a new subsequent, logical, compatible form of existence, more complex, yet, simple at the same time.

Platonic Solids along with my humble suggestion, by adding a Sphere, 6 Solids in all, illustrate transformative potency of **Intelligent Design.**

Platonic Solids along with a Sphere represent byproduct of motion, interaction between compatible opposites (gravity as well). Perhaps **Micro-cosmos is parallel with Macro-cosmos**, where, de facto, in vast space dwell inanimate matter, which its equivalent and present beyond subatomic reality (energy; potential vs interactive).

Someone have asked about essential ingredients of the Univ3rse. Well, according to the Paramount Law of Transformation, Biological Blueprint of the Universe, 6 Platonic Solids represent system, based on interconnected (Quantum) mechanism, Intelligent Design, I have indicated in previous articles.

Sphere along with Platonic Solids are classic representations of quantified projection of progression: Sphere, Elipse, Rectangle, Square, Triangle … .

Quantum (the essence of the Universe):
* singularity dwells in whole, yet, all dwell in singularity = I am in you, you are in me … .

Note:
     I am puzzled, how infinitely potent minds (quantified, sequential) can believe in finite phenomenon of time … .

The Paramount Law of Transformation.
Biological Blueprint of the Universe.
Space Program sinc3 1452.
Mass. Universal Composition.

Mass: tangible property of the Universe, as we know it. Mass reflect, de facto, essential relation between Potential and Interactive energy, as well as approximation with regard to transition from molecular to energy.

Mass is an indication, a natural occurring definition of physical, molecular vs physical energetic phenomenon, a subsequent transition sculpting reality, visible as well as non visible, yet, the very beginning of Intelligent Design (data).

Mass; Proportions between molecular potential and energetic interactive ... .

...

Universal Composition:
Limited physical, molecular elements, energetic essential phenomenon vs unlimited/infinite derivatives; molecular as well as energies = the beauty of simplification ... .

Signed,
Intelligent Design

The Paramount Law of Transformation.
Biological Blueprint of the Universe.
Space Program sinc3 1452. Logic vs Divine Affairs … .

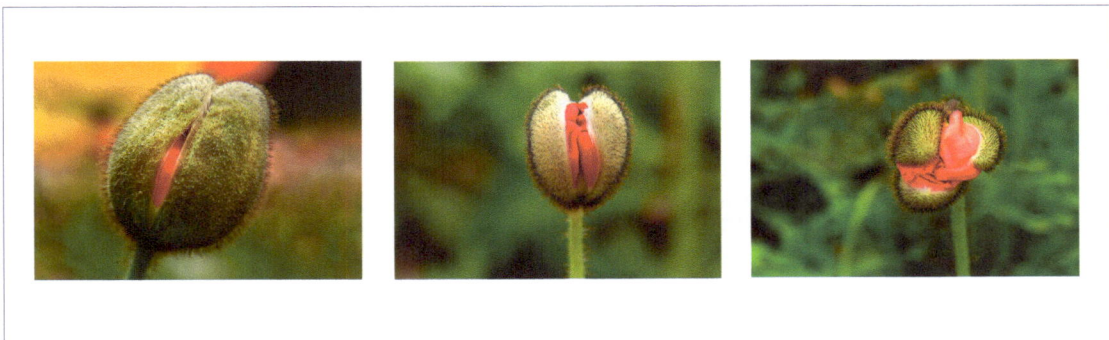

Often we hear about infinite Universe, infinitely hot, infinitely dense etc. Yet, science already made a calculated guess about the age of the Universe, as we know it : 13.7 billion years. This very number indicate finite nature of the Universe as well as the sequential, quantified chapter, where Universe will reach a stage of inanimate Space.

DNA Code indicate pre-designed biological data vs spectrum of light, electromagnetic waves, de facto, the propeller of life (DNA without spectrum of light, electromagnetic energy is useless).

* **Logic** manifest derivative projection of progression, de facto, reaction between electromagnetic waves (energy / spectrum of light)) as well as molecules.

* **Hum5n** brain reflect classic quantified system of the Univ3rsal projection of progression, de facto, exact model of transformative beauty of Int3lligent Design.

* **Human brain**, reflect perfect 3D model of the with regard to primary function, in this instance, navigation vs spectrum of light vs location rendered in 3D space. Human thought is projecting primary function as a reference point, with regard to the source of data, electromagnetic waves, spectrum of light. Once quantified brain is exposed to light its function is intensifying and vice versa. Light, spectrum, is the carrier of data, source of transformation within Universe as we know it.

* **Soul** represent a state of aware energy, essential data, definition of the perfect projection of progression, toward maximizing potential beyond basic self preservation.

* **3motions** manifest primary function, tool, based on interaction between molecules and light (since initiation sequence), electromagnetic energy, where data, through divine transformation and transitions reached a stage of higher state of projected progressions. Emotions, essential toolbox with regard to self preservation and maximizing potential toward infinity.

* **Intuition** represent projected algorithm of past experiences, de facto, tangible data vs projection of possibilities. Intuition represent sophisticated analytical tool, computation of data.

* **Sexual** function reflect, de facto, essential molecular replication/cellular computation, since initiation of the Universal sequence occurred. Truly devine Intelligent Design. Biological sexual function crossed a threshold beyond necessity.

* **Love** manifest Divine tangible algorithm in 3D space, essential projection of progression with regard to molecular world (very much social) into the higher projection of progression, in this instance, love. Love, the primary function is self preservation, through self replication, via implementation of maximization of the potential. Love is the most sophisticated tool with regard to replication of projected progression toward infinity as well as perfection.

* **Awareness** is the ultimate measurement tool with regard to Universe. If awareness assumes that Universe is infinite, than it is, yet, if awareness assumes that Universe is finite, than the space is limited and defined by the boundaries, yet, not limiting infinite character of the Universal projection of progression.

* **Something form nothing ?** It's possible. Once you have a two compatible opposites, than world begins to spin around. Computation and replication is a natural projection of progression, according to The Paramount Law of Transformation. Biological Blueprint of the Universe. Space Program sinc3 1452. Logic vs Divine Affairs … .

The Paramount Law of Transformation.
Biological Blueprint of the Universe.
Space Program since 1452. Science vs Physics vs Love ... .

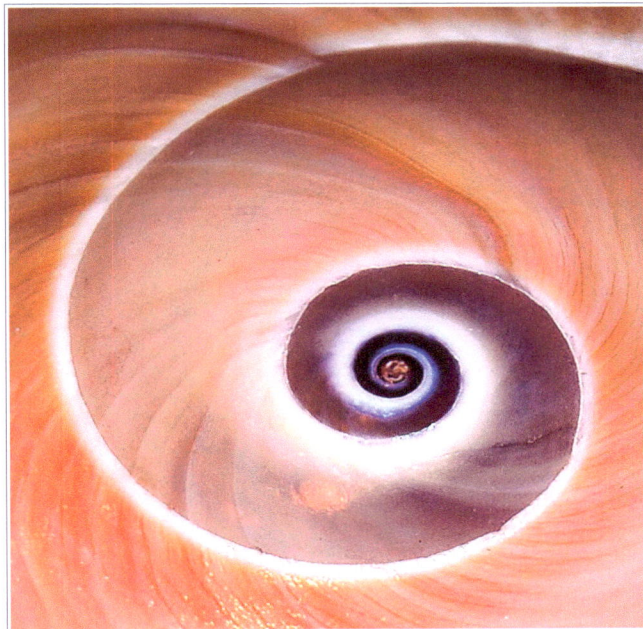

Love manifest tangible science, in this instance, projection of progression, pre-designed data to maximize potential with regard to Universal sequence of transformation.

Love, let's analyze population vs sophistication. Once you look at the biological forms of life, you will discover startling phenomenon associated with unparalleled growth, and breaking boundaries, de facto, perceptual, intellectual attainment, and certainly fulfillment in subsequent stage, with regard to social character of molecular and energetic performance.

In addition, love is accelerating perceptual, artistic expression, the very creative force, which dwells since initiation of sequence, and sure prior to Universal initiation. Tasks associated with love are often breathtaking, yet, proving practical paradigm with regard to maximizing potential.

Where love dwells, than subsequent anticipated projection of progression and accelerated pace with regard to sophistication is undeniable and attainable.

Love projects a tangible guarantee, embedded at a certain phase of sequence, a new scientific paradigm, very much divine science, which is leading sequential projection of progression toward attainment of a new frontiers, aware energy.

Love represent scientific term, a threshold allowing to reach another level in Universal sequence, this time, well defined science based on awareness, emotions and free will.

Love, anticipate tangible brilliantly designed science, based on integrated and unified forces, molecular, energies, as well as attainment of a new frontier.

Another quality of science of love is accelerated pace with regard to potential progression vs projection of progression, the essence, as of today, of sequential transition. Without the science of love, human will face progressive regress toward biological organisms, which dwell by the basic requirement of procreation, instead of maximizing potential in terms of awareness, intelligence, fulfillment, aesthetic paradigm within Universal projection of progression and obviously beyond molecular state of being.

Love is an opportunity, scientific discovery of an enormous potential with regard to energy, integrated fragmentation of a whole into whole of a quantified energy, as well as integrated power of social energy. There nothing more powerful and promising, than this science based on energy (emotions) of performance. Pure science ... .

Love is a natural physical and energetic occurrence since human body is considered an art form; biological engineering, practical solutions, de facto, in all aspects embedded within Universal sequence.

Love, to disregard scientific achievement of this phenomenon, which not only produces tangible results, de facto, positive, yet, manifest essential platform to attain something that is very unique and very human, satisfaction and purpose to build sequential quantified system in an integrated, in this very moment, transition, where all doors and gates of the Universe are open, scientific, perceptual, and energy progressing beyond known quantitative frame of aware participation, yet, very much promising.

Love, powerful energy, once compatibility is integrated... .

The Paramount Law of Transformation.
Biological Blueprint of the Universe.
Space Program sinc3 1452.

* Energy: High Velocity Plasma … .

Universal system is oscillating within transition, very simple idea, indeed;
* Energy (high velocity plasma) is gradually transformed into formation of progressively heavier molecules, which produce an energy, and an energy is transformed into molecules, and molecules produce an energy, and energy is producing molecules; and this is wonderful Divine music of Multiverse … .

Yet, some fraction of aware biological aware order is progressing into aware energy, beyond anticipated, as of today spectrum of energetic projection of progression, oscillating beyond limitation of more dense molecular energy.

…

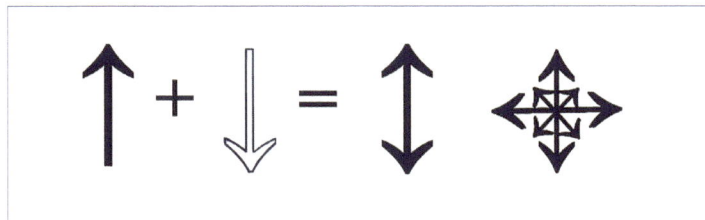

Graphical representation of the Universal Initiation Sequence
Hot Light + Cold Light = High Velocity Plasma
Cold Light: inanimate space/matter prior to Initiation sequence.

The Paramount Law of Transformation.
Biological Blueprint of the Universe.
Space Program sinc3 1452.

Univ3rse vs Gravity ... .

Gravity is associated with density of molecules, yet, is less bounded by gravitational forces, while energy becomes more refined; transition from potential energy into kinetic energy.

Gravity is produced by compatible opposites (spectrum of light vs cold light), subsequent motion, as well as density of matter compatible with regard to molecular and energetic data.

It seems that there is a fine difference between gravity and vortex of matter, yet, very much identical, while projecting its molecular and energetic imprint in 3D Space. The tipping point is a threshold with regard to intensity of radiation. Vortex of Matter is kind of Gravity on steroids ... .

...

The Paramount Law of Transformation. Biological Blueprint of the Universe. Space Program sinc3 1452.

Univ3rse vs Vortex of Matter ... .

Vortex of Matter manifest projection of progression formed by extremities; intense, powerful radiation, spectrum of light vs cold light (inanimate matter). In this very moment, Vortex of Matter is formed once a threshold between two compatible opposites allows to form molecular vortex, which transform large quantity of data into molecules and energy. Same phenomenon exists on Earth as well.

Formation of vortex of matter is somehow "breaking the symmetry", yet, more precise description would be transition from one symmetry into a new progression of projection, symmetry, de facto. In terms of graphical/geometric, mathematical illustration, „breaking symmetr y" is easily rendered through transition from one geometrical object into a new geometrical object (illustration on page 85) , while first is the beginning of a subsequent projection of progression, which is the beginning of the next projection of progression ... .

The Paramount Law of Transformation.
Biological Blueprint of the Universe.
Space Program sinc3 1452.

Space Bridges … .

What is propelling the Universe ?

* Intelligent Design … .

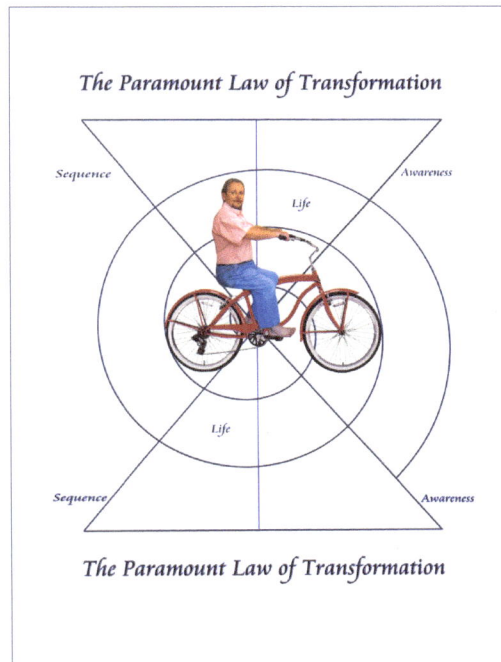

The Paramount Law of Transformation

Sequence

Awareness

Life

Life

Sequence

Awareness

The Paramount Law of Transformation

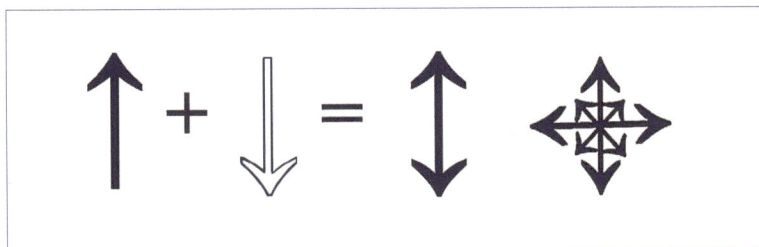

Hot Light + Cold Light = High Velocity Plasma

<h1 style="text-align:center">Cosmic Bridges … .</h1>

As I have indicated, and proved over the course of my book, Universe is profoundly intelligent, while interacting with the source of transformation, spectrum of light vs inanimate matter. Data associated with **Universal Transformation**, are **Space Bridges**, de facto, **Resonant Waves of Data,** propelled through Intelligent Design in a direction anticipated, as well as a path to maximize potential in an optimal sequential progression possible.

<h2 style="text-align:center">Eight Minutes Universal Bridge;</h2>

With regard to Earth, Human civilization, essential data extended by the Divine force of illumination, represent **8 minutes between Sun and the Earth**, yet, Earth is being fertilized also by other sources of data from space, via cosmic compatible resonant waves … .

<h2 style="text-align:center">String Theory …</h2>

String Theory applies to the strict path, precisely designed, with regard to anticipated projection of progression. It is, de facto, precise science of Resonant Waves. String Theory, as such, is non existent, yet, could be translated as a vector within quantified (fragmented) Universal system, refined and precise gyroscope, aiming to achieve maximum benefits, yet, with least effort.

In terms of Biological equivalency, human races perfectly illustrate Divine plan toward diversity, yet, closely bounded within strict, and logical sequential progression.

…

<h2 style="text-align:center">Horizon</h2>

Horizon represent a point of awareness, due to the Paramount Law of Transformation – everything is in motion, yet, static horizon contradict kinetic nature of the Universe, including Multiverse … .

Perhaps we shall ask; at which stage of quantified transition is particular sequence. Yet, horizon manifest poetic, and philosophical notion of the past, quite elegant. Projection of Progression vs Music.

The Paramount Law of Transformation. Biological Blueprint of the Universe. Space Program sinc3 1452. Space Bridges …

.

Human and Universe as One Interconnected Organism. Simplicity vs Complexity

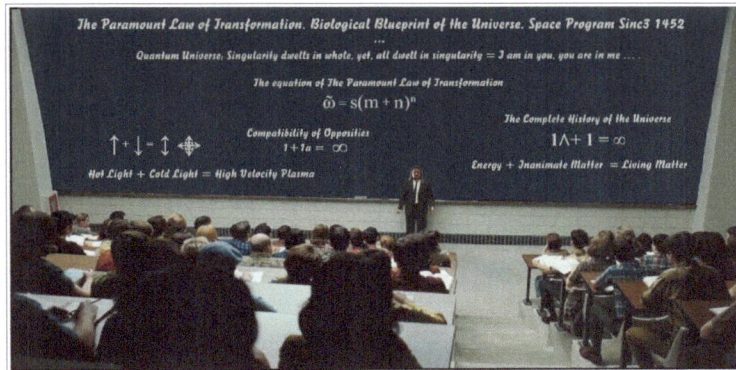

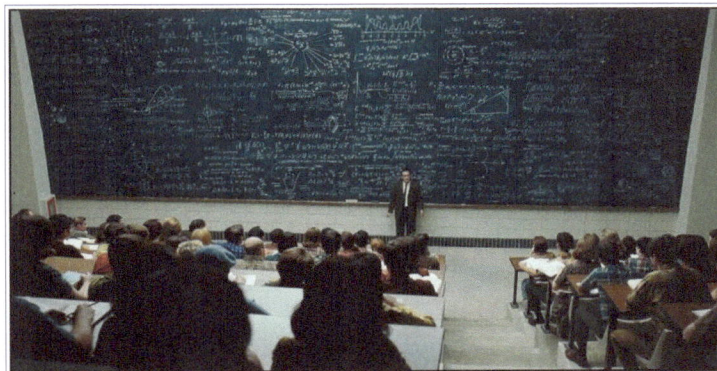

The essence; fundamental paradigm of the Universe, as we know it;

* compatible of opposites, which is emerging in every algebraic as well as geometrical sequence3 … .

Universe resemble all stages of human development from initiation sequence through growth and subsequent stages. Yet, human is far more profoundly wired with the Universe than it is anticipated as of today, except the notion of the interconnected world is already explored by Asian cultures.

Human collective energy is shaping Universe, same as Universe miraculously shaped all stages of existence up to the point where living, aware energy molecules emerged in vast space.

If apparent interconnection, wired Universe is performing via projection of progression than human performance is fundamental with regard to performance of the Universe. This is yet to be determined by science, yet, by using Human as a Biological Blueprint of the Universe is, de facto, proving that the entire sequence is simple and divinely efficient.

Compatible Opposites suggest that human is shaping universe and Universe is miraculously shaping human, yet, aware energy represent the most sophisticated projection of progression to date, except mechanism, which initiated Universal Sequence.

* Initiation of the Universal Sequence Creative Force
* Projection of Progression
* Aware molecules and energy = Human
* Molecules = Energi3s

The Paramount Law of Transformation.
Biological Blueprint of the Universe.
Space Program sinc3 1452.
Space Bridges – Red … .

Why human perception is so profoundly attached to red color, an explanation of this phenomenon with regard to perceptual awareness is obviously, gracefully embedded in science:

* red represent visible spectrum of light
* visible spectrum of light is potent and allows procreation
* red, visible spectrum if light is also indicating that human brain, operating system (gyroscope) is navigating and coupled with light, in this instance red light
* red color=red spectrum of electromagnetic waves (the shift) is profoundly embedded in biological existence, as well as orientation with regard to the source of transformation
* red is sexy because represent the essence of survival as well as associated pleasure
* human brain is the antenna while detecting electromagnetic data.

The formula/coordinates of transformation and pleasure;

Spectral coordinates;
Wavelength ~620–740[1][2] nm
Frequency ~480–400 THz
About these coordinates Color coordinates
Hex triplet #FF0000
sRGBB (r, g, b) (255, 0, 0)
Source X11
B: Normalized to [0–255] (byte)

Rendering of the spectrum of visible light (sRGB);

Color Wavelength Frequency Photon energy
* violet 380–450 nm 668–789 THz 2.75–3.26 eV
* blue 450–495 nm 606–668 THz 2.50–2.75 eV
* green 495–570 nm 526–606 THz 2.17–2.50 eV
* yellow 570–590 nm 508–526 THz 2.10–2.17 eV
* orange 590–620 nm 484–508 THz 2.00–2.10 eV
* red 620–750 nm 400–484 THz 1.65–2.00 eV
(Wikipedia)

The spectrum is continuous, with no clear boundaries between one color and the next. Quantum; the essence of the Universe:
     * singularity dwells in whole, yet, all dwell in singularity = I am in you, you are in me … .

The beauty of the spectrum of light is a seamless transition from wavelength to wavelength, yet, all interconnected into synthesized projection of progression via data tube = the Sun … .

The Paramount Law of Transformation.
Biological Blueprint of the Universe.
Space Program sinc3 1452.
Space Bridges vs Flowers … .

Why flowers know when to bloom ?

Because they are exposed to light, energy of transformation, data embedded, packed via Intelligent Design, and when time is right, electromagnetic waves unpack its potential … .

. . .

How the universe is built ?

Universe is an idea projected into the blueprint of molecular progression, and apparently projected within an idea, subsequently images appear in Space, yet, this is just a part of the story, because we are projecting back singular, as well as collective vision within inner projection of progression, that's why is so essential to project the finest images man is able to compose in hearth and mind, as well as social interactions … .

. . .

Earth is a living organism and is in motion, orbital, around its axis, as well as within our Galaxy, and apparently the Universe. Due to the organic character of our Planet, Earth is shifting its shape and is never in a fixed state … .

. . .

Whenever you experience a bad hair day, that's because interplanetary wind is interfering with well combed hair style – we are traveling in space and the speed is not only breathtaking, yet, Universal engineering turns and twists our Planet along with humanity in multi directional vectors of motion; probably 5 in all; around its Axis, Orbital, Solar System, Galaxy-Milky Way, Universe – wow … .

. . .

Universe loves motion, indeed, and is never tired of new dancing moves … .

Universal projection of progression is strangely pragmatic,, de facto invented via Intelligent Design, love, to extend existent projection into infinity, along with pleasure, virtuous emotions embedded within compatible opposites, where packing and unpacking of data is so appealing … … . Human, as well as Universe is packing and unpacking its data, what a journey. Tomorrow, in the morning, Universe will unpack its gift of life as well as renewal, and the next gift and the next day, and the next … .

. . .

Looking for the formula of love? Approximation dwell in the spectrum of light, yet, translate this formula into sounds, and surely Golden Ratio will sing in your heart and mind … .

The Paramount Law of Transformation.
Biological Blueprint of the Universe.
Space Program since 1452 ... .
Projection of Progression
Necessity vs Pleasure ... .

Universe progressed from necessity to pleasure,and this is a truly Divine quality ... . Once self awareness emerges along with pleasure, than a profound projection of progression threshold is being crossed ... .

The purpose of awareness, as well as pleasure is to extend existence beyond boundaries of necessity and subsequently procreation. In this very blueprint Divine is creating Gods in the making, yet, as a mirror reflection to maximize its potential, and at the same time minimizing risks, that's why ethical standards were invented, as a safe path toward infinity ... .

The Paramount Law of Transformation.
Biological Blueprint of the Universe.
Space Program since 1452 … .

Journey within Quantified Projection of Progression

Same man appears twice in the image – practical sequential travel. This is a proof that sequence is quantified, yet, our path is like a train (string theory) we can access any sequential projection at will.

Image provide practical illustration how to access any sequence, yet, there is only one, according to Quantum Physics.

Quantum Universe:
* singularity dwells in whole, yet, all dwell in singularity = I am in you, you are in me … .

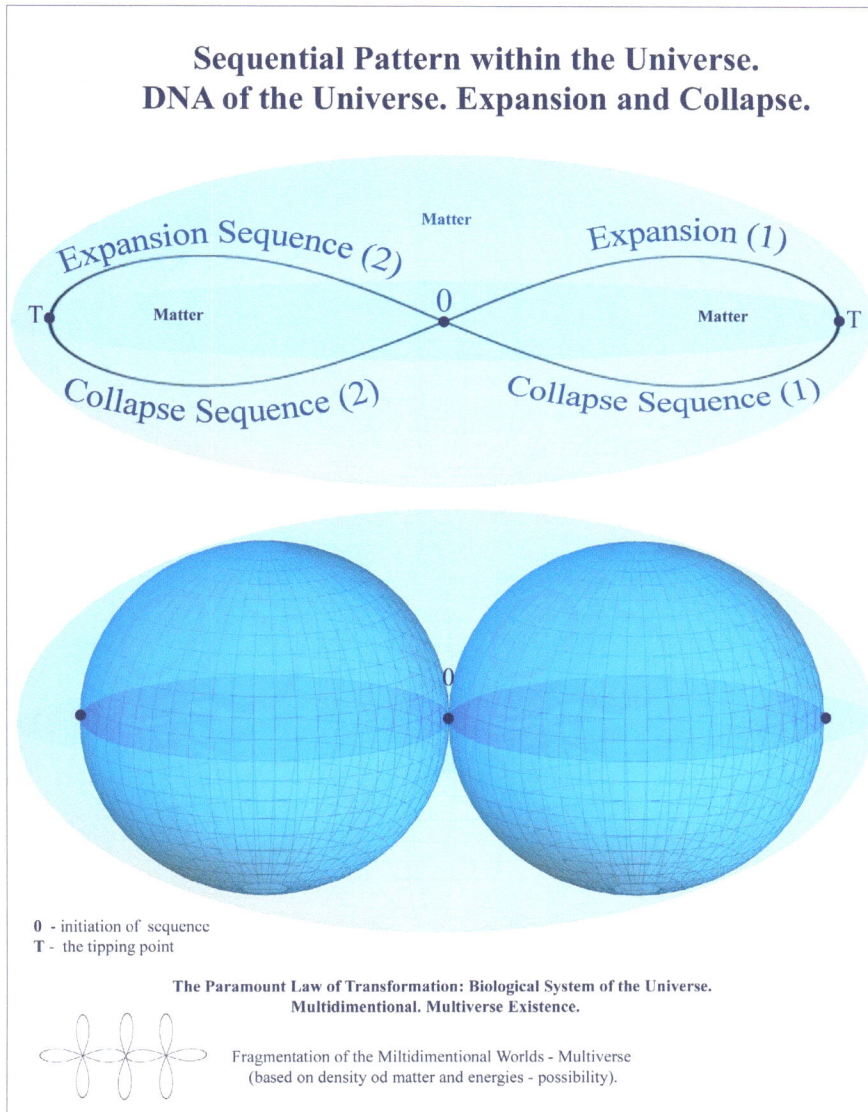

Sequential Pattern within the Universe.
DNA of the Universe. Expansion and Collapse.

Matter

Expansion Sequence (2)   Expansion (1)

T   Matter   0   Matter   T

Collapse Sequence (2)   Collapse Sequence (1)

0

0 - initiation of sequence
T - the tipping point

The Paramount Law of Transformation: Biological System of the Universe.
Multidimentional. Multiverse Existence.

Fragmentation of the Miltidimentional Worlds - Multiverse
(based on density od matter and energies - possibility).

The Paramount Law of Transformation.
B;ological Blueprint of the Universe.
Spac3 Program Since 1452.
Space Bridges ? Y3s … .

Quantified Unity

vs

Logic5l Projection of Progression of Shap3s Der1ved from Univ3rsal Motion … .

...

Universal Lesson; How to Compose the Most from the Least … .

...

Progressive Segmentation Sequence Project within Universe … .

3D Eloquence of Motion; Spiral, Sphere, Elliptical, Triangular Shapes, Square, Rectangle... . Today I would like to tackle with a subject I already explored in previous articles, yet, requiring addition comment.

How world was cr3ated... . Well, motion comes to mind, simplicity, as well as a reference point, with regard to the source of transformation, electromagnetic spectrum, to say the very least. Plasma is cooling down, and begins to turn in circular motion;

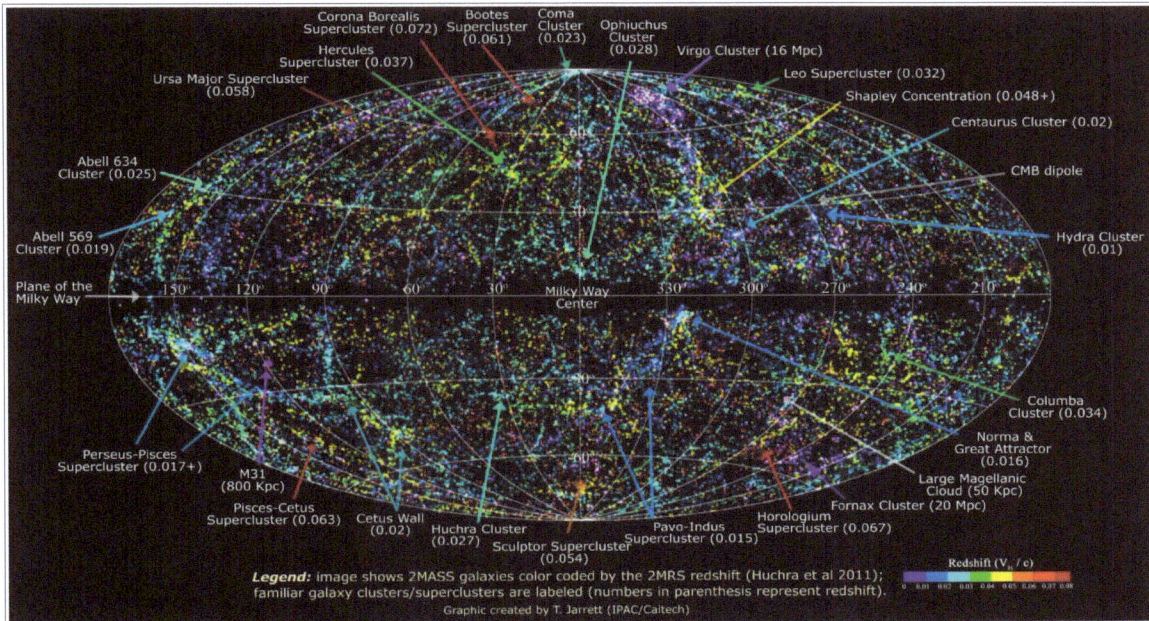

Universe from Plasma to Densiti3s ... .

The most efficient way to illustrate miraculous formation of our Universe, is geometry, because our fantastic minds operate in 3D.

Initiation sequence manifest, de facto, motion, yet, vortex is the essence of all good things Divine has on His blueprint board;

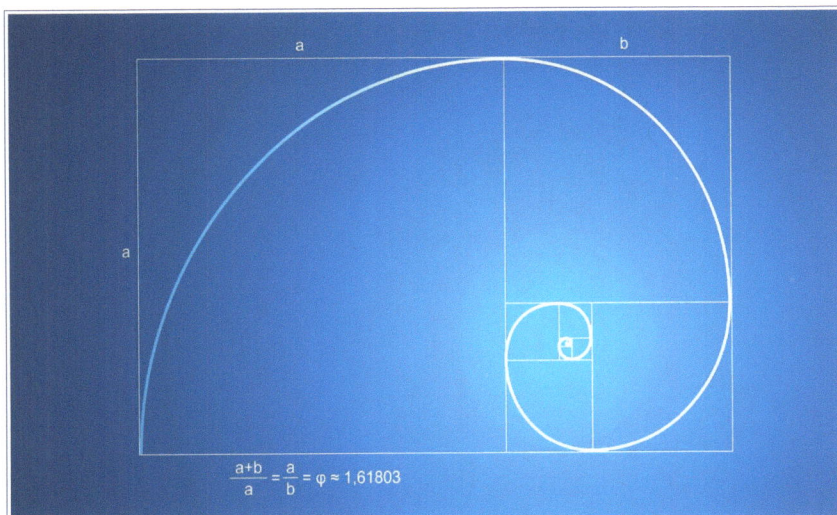

$$\frac{a+b}{a} = \frac{a}{b} = \varphi \approx 1,61803$$

Once motion molded a spiral, subsequent sphere/circle emerged;

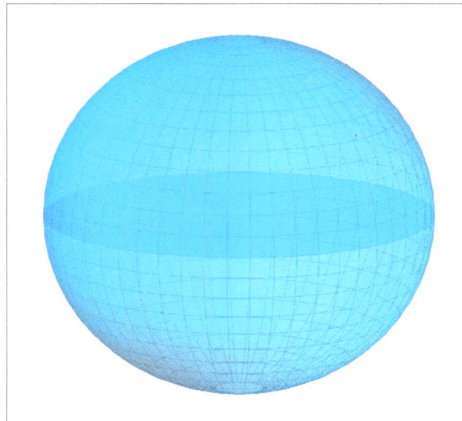

From sphere/circular elliptical shape derived orbits;

Formation of heavenly bodies begin to take shape, and their corresponding positions formed a Tri5ngle, Squar3, Rectangle, Trapezoid, and subsequent spherical forms, which were, and are replicated throughout the Universe ... .

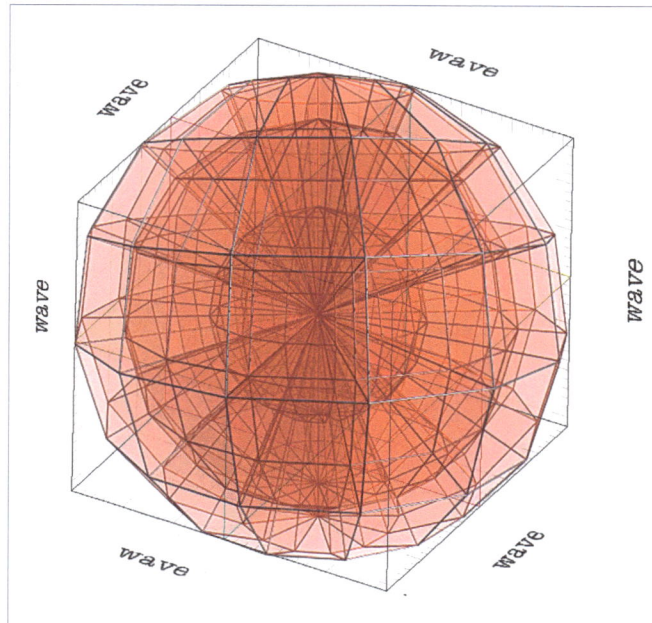

Logic5l Projection of Progression of Shapes is Der1ved from Univ3rsal Moton is compatible with progression of Pithagorean Theorem, yet, the notion of Theorem is taken to the very initiation of Universal sequence. In addition, Pythagorean Theorem is compatible with Golden Ratio … .

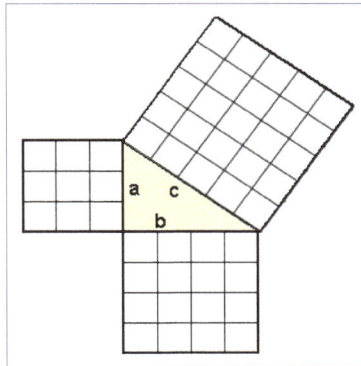

…

Look, Gravity; Human was Assembled via Gravitation Force (assembly facility), de facto, essential tool of the Creation … .

Biological Blueprint of the Universe is projecting identical properties within Universe, for molecules, living unaware, as well as living, aware;

Compatible Opposites (man, woman) = "assembling" a new manifestation of existence. This phenomenon represent gravitational paradigm allowing molecular composition, procreation … .

Man dwells in Unified composition of internal, as well as external projections of progression;
Two worlds, two projections in one.
This notion is also compatible with molecular, energetic, as well as psychological, social interactions.

The Paramount Law of Transformation
Universal Progression of Compatible Opposites

Complete History of the Universe vs Biological Blueprint of the Universe

$$1\wedge + 1 = \infty$$

Energy + Inanimate Matter = Living Matter
(Hot Light + Cold Light = Initiation of Perpetual Universal Progression)

Initiation of the Universal Sequence through Compatible Opposites

Transition of Living Matter into Aware Matter and subsequently
into the Aware Energy.

Biological Blueprint of the Universe

The Paramount Law of Transformation.
B;ological Blueprint of the Universe.
Spac3 Program Since 1452. Space Bridges ? Y3s … .

What dwells beyond atoms, molecules, subatomic particles-energies ?

vs

Plasma … .

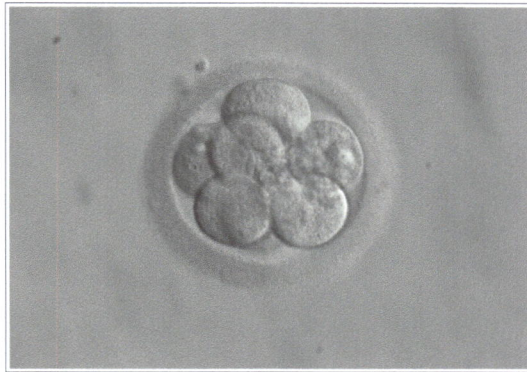

Embryo; 8 Cells.
Molecular/cellular projections of progression.
Embryo-Human Cells Nucleus

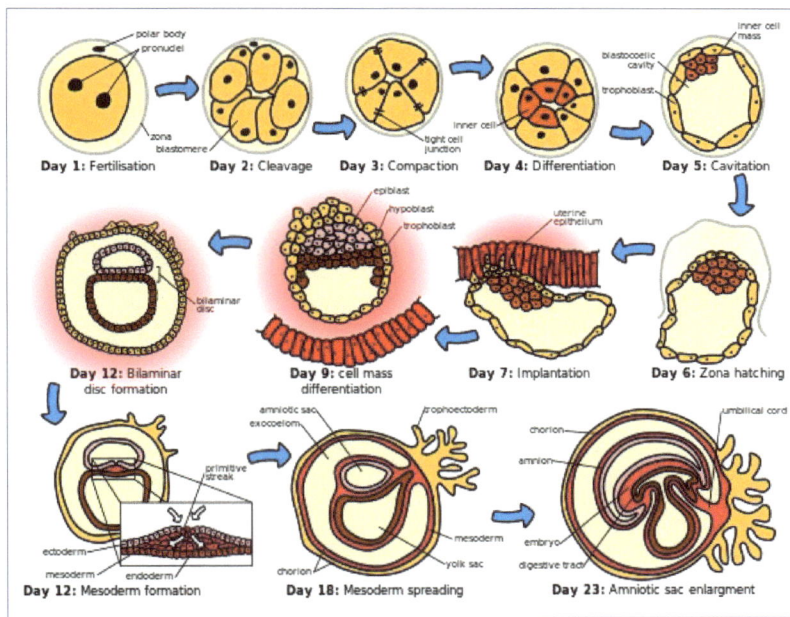

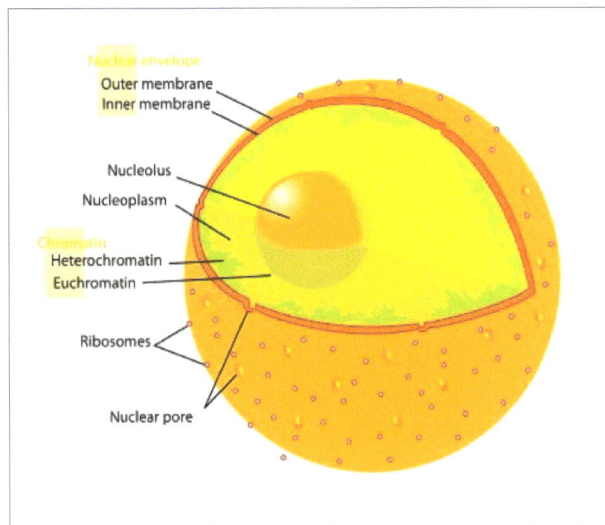

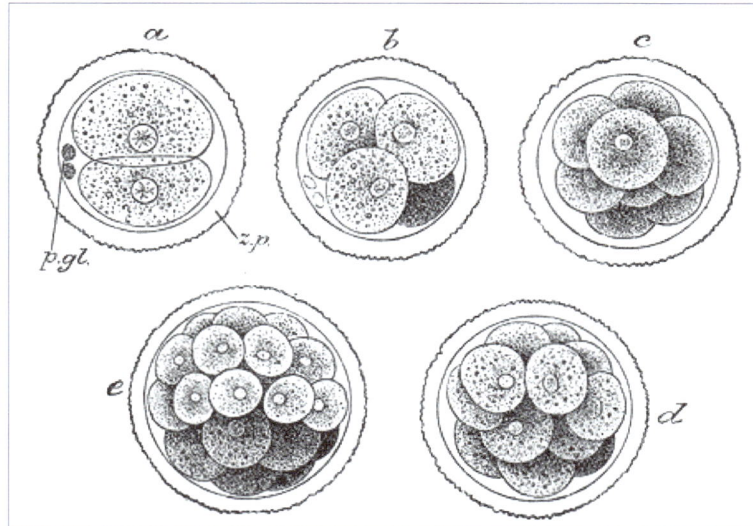

Embryo

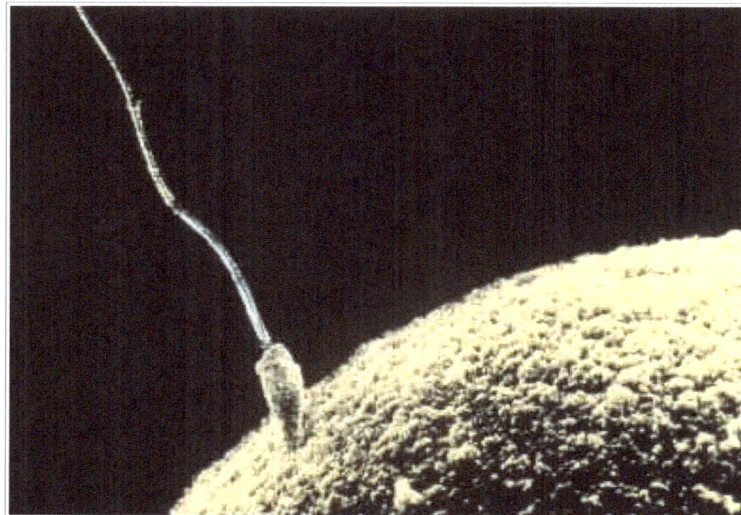

Embryo and Sperm; creation of the new, subsequent projection of progression.
(Images credit: Wikipedia)

Cytoplasm may appear to have no form or structure, it is actually highly organized (Wikipedia).\ Yet, cytoplasm manifest corresponding quality with regard to structure of data within which subatomic particles/energies dwell. We know that everything within Universe is energy, either potential or kinetic, de facto, a potent data, able to transfrom, engineered and destined for transitions.

Nearly Neutral Background Plasma, beyond subatomic projection is able to „borrow/provide" an energy progressing from potential to kinetic; identical molecular mechanism is observable with regard to Protons. In addition identical property define human interactions within singular interactions, as well as collective (exchange of an energy). The source of tthis interactions dwell in subatomic projections.

The Paramount Law of Transformation.
B;ological Blueprint of the Universe.
Spac3 Program Since 1452.
Space Bridges ? Y3s … .

Mirroring vs Universal Acoustics
Molecular Echoeing vs Sound
Quantum Tunneling
Univ3rsal Enthusi5sm

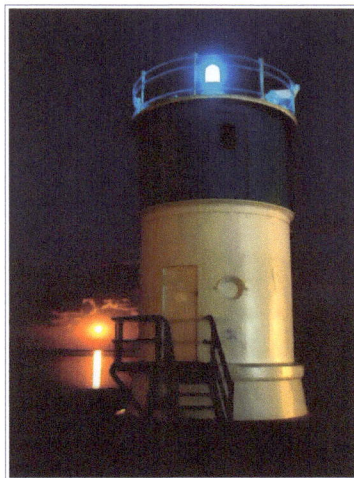

Mirroring vs Universal Acoustics;

I wrote 10 years ago, on my website while living in the USA about Acoustics, yet, this issue still require attention and further exploration.

Often we hear about **Mirroring of Universe**, yet, as much as this notion is attractive, and possible, Universal Acoustics describe the notion of parallel universe more eloquently, due to the fact that is performing within 3D Architecture.

Universe represent, and manifest biological kinetics, as well as properties, yet, defined profoundly by cellular automation (greatly explored by Mr. Wolfram).

Acoustics are resonant waves, very much present throughout the Universe; noise, sound, by the very nature, represent higher order of sophistication, up to the point, where necessity progresses into projection of aesthetic paradigm, very much embedded within all spectrum of data, energy, molecules, energy, aware energy. Acoustical projection of progression is than formulating subsequent apparent cells of the multiversal organism;

Quantum Universe:
* singularity dwells in whole, yet, all dwell in singularity = I am in you, you are in me ...

This is my favorite quote, because is illustrating **Intelligent Design**, brilliancy of the **Biological Blueprint of the Universe**, simplicity in complexity ... .

We are an integral part of the whole organism, which evolve still, through Acoustical properties of the resonant waves, echoing, yet, different, and very much more sophisticated.

System is brilliantly simple; as molecular data evolve reaching stages of sophistication, than Universe, as we know it, does replicate the very same properties, while transfering Acoustical echoes, imprint of data, into the subsequent architectural space, where next stage of cellular automation, de facto, energy is evolving beyond present frame of awareness ... . Yet, we can touch it, because system is replicated according to the same principles of the design ... .

This superbly designed phenomenon is not only biologically observable, proven, but manifest its quality within singular organism, self contained, as well as social, also self contained, yet, all is interwoven within each structure. This is called **Infinity**, and **Universal Acoustics** play essential role with regard to transfer of data within space architecture, which, unlike human made structures, is able to change its shape, size, adopt and evolve. This is a **Universal Architecture** based on Biological Blueprint of the Universe. Human, what a great design, Divine, simple, yet, efficient in every drop of creative data embedded in all components within **The Paramount Law of Transformation... .**

Divine, there is no other descriptive eloquence, which would illustrate projection of progression within design, yet, Divine Projection replicates itself, fit perfectly within scientific frame of anticipation, as well as artistically motivated progression.

The proces of molecular assimilation, **Molecular Echoeing**, convergence is also profoundly manifested in human environment, relations, where humans not only assimilate, but rapidly blend physically.

...

Sound is associated with motion, and motion, by the definition, represent higher order of prgression, embedded within **3D space**. Sound is parallel with 3D progression, which I explored previously; from necessity to pleasure. Voilà, Vooilà, Voooilà ... .

...

Quantum Tunneling refer to transition with regard to density, which allows to achieve similar stage and result, or identical, without excessive amount of energy, but reduction of density to reach the other side of the hill; similar solution is observable in properties of enzymes; reduction of temperature, yet, achieving anticipated, essential results.

Quantum Tunneling is identical with **Universal Enthusiasm,** as well as **Biological Aware Enthusiasm**; man is composed from a very intelligent, intellectually potent ingredients, molecules, and energies, yet, this intelligence is projected within singular properties, as well as social interactions ... . Universe is a very creative organism via Divine replication, yet, creativity is reinforced with enthusiasm; pleasure represent a higher form of enthusiasm ... .

...

Quantum Theory of Light refer to data, and subsequently data within data-spectrum; Quantum Universe once again. Spectrum of light, is generated via motion.

...

The Paramount Law of Transformation. B;ological Blueprint of the Universe. Spac3 Program Since 1452.

Quantum Watch
En3rgy; all we were, all we are, all we will be … .

If subatomic particles are able to borrow energy or transition from one subatomic state to another (photon vs electron vs positron), than everything that we surrounded by is an energy, either transformable from potential or/to kinetic. Another phenomenon with regard to borrowing energy is that energy is embedded within inanimate, as well as kinetic physical manifestations, states, where a threshold from inanimate/potential and kinetic is defined by the interaction of compatible opposites. Energy is always available, if needed … .

Subsequently the entire model with regard to Universe is an energy, compatible opposites, which create a matter.

En3rgy; all we were, all we are , all we will be … .
Yet, matter is created from energy, compatible opposites .

Quantum Cycles Watch
Quantum; the essence of the Universe :
* singularity dwells in whole, yet, all dwell in singularity = I am in you, you are in me … .

Wheels are turning by an energy; spectrum of light from subatomic, through bicycle, Universe … .
What is turning the wheels, energy. Wjhat is turning, formulating energy, Intelligent Design ... .

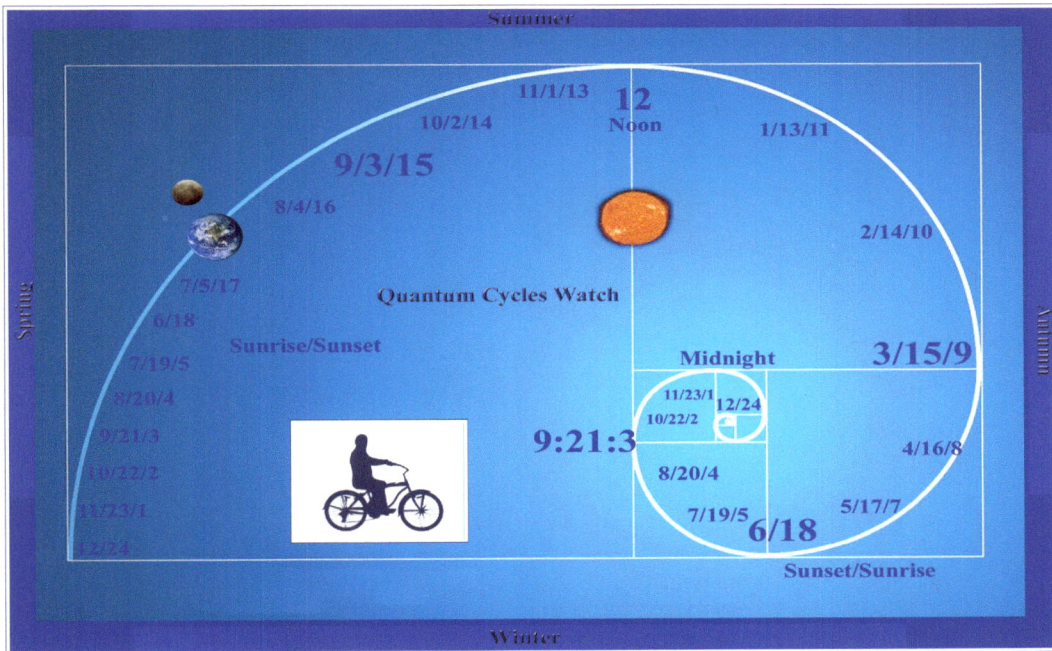

The Paramount Law of Transformation
Milky Way Design
Pierwotne Prawo Transformacji
„Droga Mleczna" Projekt

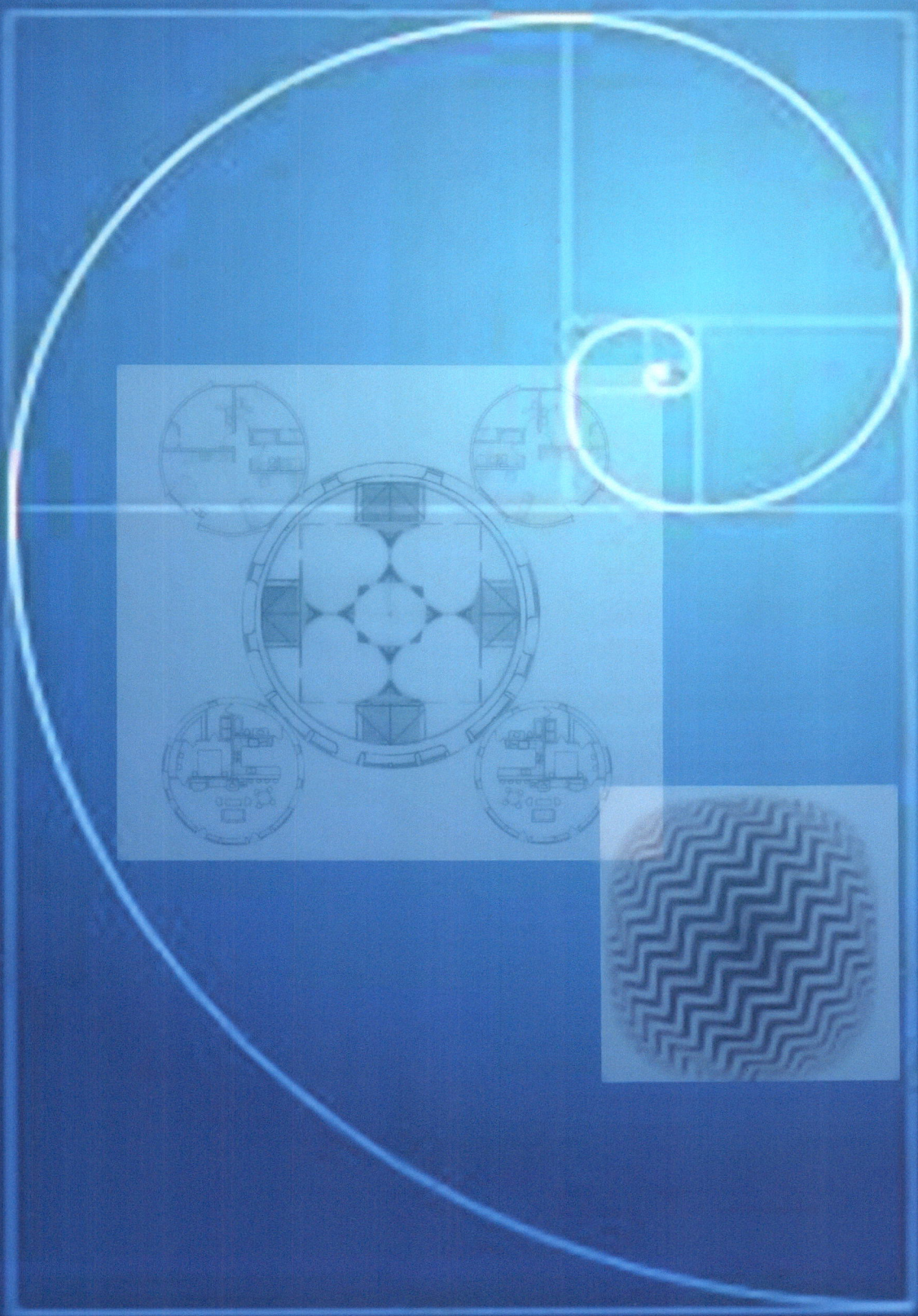

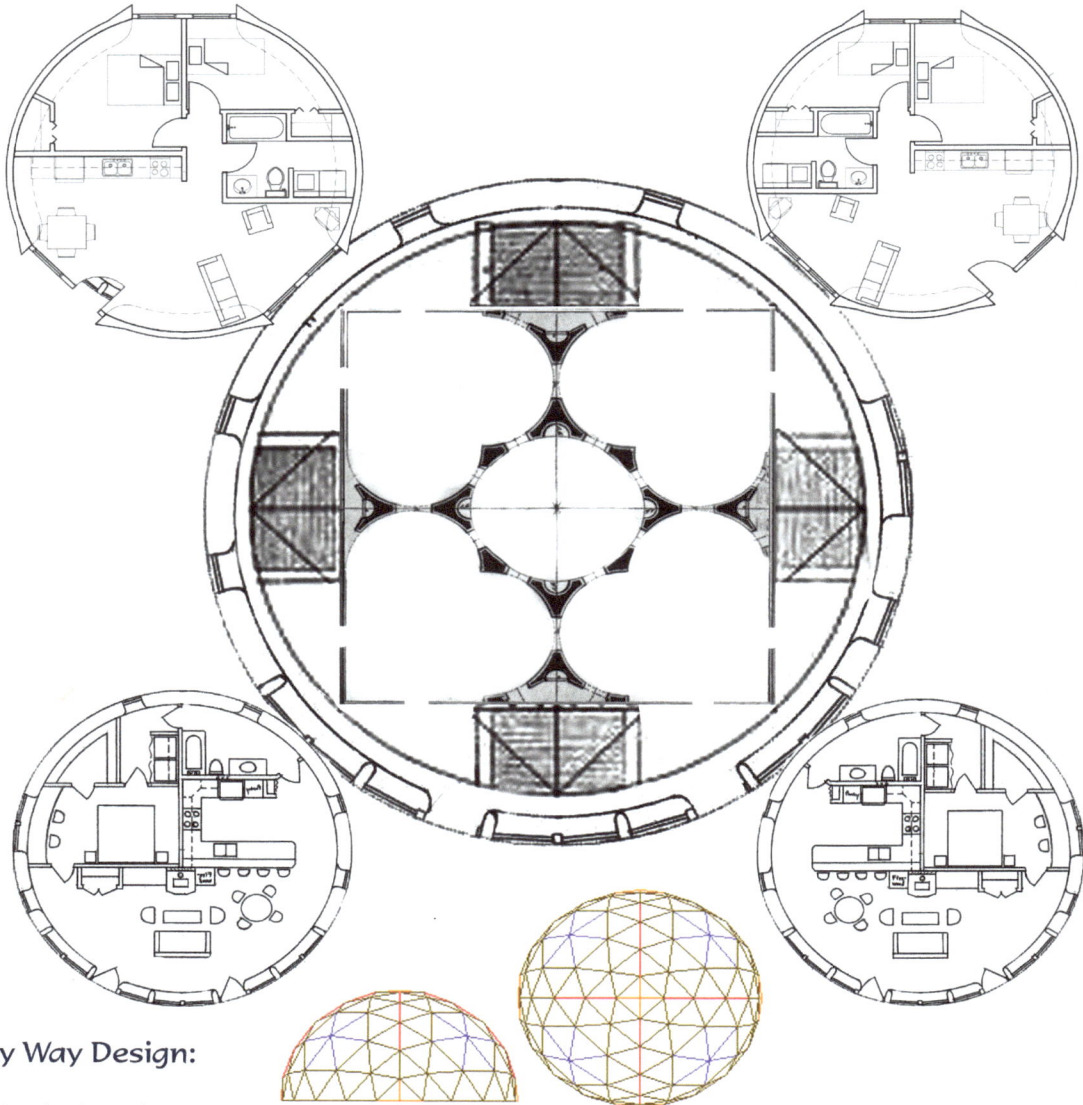

Milky Way Design:

1. Spherical Motion
2. Rotation around its axis and revolution
3. Glass dome (solar panels, generators of electricity).

 Design is based on spherical motion of Earth around its axis as well as revolution around the sun. In addition, motion is achieved through electrical power from the solar energy. Motion generate electrical power. Entire structure is build from eco friendly materials, solar panels which generate electricity. The Paramount Law of Transformation "Milky Way" design reflects motion of the Universe, profoundly common, profoundly efficient, profoundly elegant. Design provide aesthetic view nearly infinite variety of angels in terms of vision vs light, vs sequence. "Milky Way" design is one of the kind, and can be adopted in variety of buildings including skyscrapers. Just imagine the view, every morning, noon, evening and night.

The Paramount Law of Transformation Design.

All Rights Resreved 2004/2015   Tel. 536 508 394   www.scribd.com/mjw23   Marek „Mark" J. Wagner   www.zhibit.org/mjw23   E-mail: q7q7mark@gmail

230

www.ingramcontent.com/pod-product-compliance
Lightning Source LLC
Chambersburg PA
CBHW041950220326
41599CB00004BA/91